Why *Space* Has *Three Dimensions?*

A Journey from Qubits to Cosmos

Anvay
Sheolikar

Book Cover Photo Credit: X-ray: NASA/CXC/U.Mich./S.Oey, IR: NASA/JPL, Optical: ESO/WFI/2.2-m

First edition

To Baba Aai, Ajooba Aaji, Aayush dada, Yogesh kaka, Anuj dada, Mahesh kaka.
And to the Humans from Earth!

Contents

Commencement iii

I The Question

1 Six Numbers 3
2 Anthropic Trap 22
3 The Argument Ahead 35

II The Cracks

4 Geometry of Gravity 43
5 Black Hole Revolution 57
6 Holographic Clue 69

III Architecture of Space

7 Quantum Connection 83
8 The Breakthrough 95
9 Building Space from Qubits 104
10 Why Three Dimensions? 114
11 Emergence 138

IV The Other Dimension

12 The Fourth Dimension 159
13 Dynamics Without Time 170

14 Five Paths to One Answer 182

V The Lab Report

15 Spacetime on a Chip 199
16 Gravitational Waves as Tests 226
17 Solar System Laboratory 241
18 The Planck Scale 252

VI What Follows

19 What Else Emerges 267
20 Engineering Spacetime 280
21 The Philosophical Turn 294
22 SPACE IS A VERB, NOT A NOUN 310

Notes 320
About the Author 329

Commencement

Sixty-six million years ago, a colossal object, roughly ten kilometres wide, collided with Earth. It struck the Yucatán Peninsula with a force akin to billions of nuclear explosions. The aftermath triggered a storm of dust and sulphur aerosols that spiralled into the atmosphere, suffocating sunlight for months. Without light, photosynthesis ceased. And the entire food chain collapsed in disarray across the planet. In just a few short years, a staggering seventy-five percent of all species perished, including every non-avian dinosaur. The dinosaurs were blissfully unaware of their impending doom. Unbeknown to them, an asteroid hurtled through the dark void of space. They couldn't lift their gaze to see the sky's warning signs because deciphering orbital mechanics was beyond their grasp, as was the idea of extinction. Bounded by Earth's surface, they stood powerless against the universe's violent whims. Their ignorance protected them from knowing that such knowledge was even possible. The same celestial mechanics that guide our modern spacecraft today influenced the asteroid's course. In principle, it was calculable, provided one had the right observational data and computational tools. The dinosaurs roamed majestically. Yet, why didn't they evolve cognitive prowess? That mystery must bide its time. What they did, however, was play the role of active spectators in a grand cosmic show without any pretense.

We humans from Earth are unique. Not only due to our intelligence, but because we have the ability to extend our senses beyond what is biological. We alone, among all creatures that have drawn breath upon this planet, have turned our instruments upon the void and demanded it answer. It is audacity without precedent in the natural world.

This book explores the concept of comprehension and its importance as we address the fundamental question: *Why Space Has Three Dimensions?*

Fast forward sixty-six million years to 23 August 2023, at 12:33 UTC; everything changed. The fabric of our relationship with the cosmos was irrevocably altered. In that single moment, humans almost found a new connection to the universe that cradles us.

A spacecraft, nearly 384,400 kilometres from our home planet, initiated its final descent at thirty kilometres altitude. The Vikram Lander—India's second landing attempt and third lunar mission—fired up engines dormant for four days. Racing down at about 1.7 kilometres per second, it aimed for a pristine location. This lunar patch, mapped from above, had yet to experience a human essence, and no nation had ever touched it. The target was located at 69.37 degrees South latitude and 32.35 degrees East longitude, between two ancient lunar craters: Manzinus C and Simpelius N. This unique region experiences sunlight filtering through at slanted angles during the lunar day, while nearby craters cast a permanent darkness. A zone untouched by successful landings. Just days earlier, Russia's Luna-25 attempted to brave this terrain but tragically crashed during its descent sequence. And thus, the measurement begins with *light*. The lander does not have wings, as there is no air to generate lift. It does not have wheels, since there is no road beneath it. The Vikram lander descends towards a surface it cannot see, guided by instruments and mathematics. It slows itself down using engines that burn fuel into exhaust, pushing against the vacuum of space. The spacecraft ignites four 800-newton thrust throttleable engines, which burn monomethylhydrazine as fuel and mixed oxides of nitrogen tetroxide as oxidizer. This pushes against the vacuum and decelerates the spacecraft's downward velocity.

The altitude is 1,700 metres. The Vikram lander is descending at a velocity of 60 metres per second. The spacecraft's computer processes data from a laser that measures the distance to the ground, radar that measures vertical velocity, and accelerometers that measure the spacecraft's motion. Algorithms feed this data to compare the spacecraft's current state with its desired state and calculate corrections. This sequence—measure, calculate, correct—runs continuously. Hundreds of times per minute, the spacecraft probes its status: Where am I? What must I change? Where should I be? With each question, it shifts direction, making precise adjustments to stay on trajectory.

At an altitude of 800 metres, the countdown begins. With a descent rate of 30 metres per second, the pressure intensifies. Computers are buzzing, recalculating the shrinking window of opportunity without pause. Fuel is being consumed at a deliberate pace. The margin for error is meagre almost nonexistent. Burn too much fuel correcting the trajectory, and the landing becomes uncertain. Fail to correct enough, and the spacecraft will greet the ground.

At an altitude of 400 metres, the guidance system slows the speed to 30 metres per second. As the surface becomes clearer, rocky terrain and craters come into view. The guidance system uses cameras to compare live images with its map, working diligently to gather data.

At an altitude of 100 metres and a descent velocity of 5 metres per second, it now hovers close enough to spot individual boulders.

At an altitude of 30 metres and a descent velocity of 2 metres per second, fuel is almost empty. If it doesn't land soon, the spacecraft will tumble.

Finally, at an altitude of 10 metres, the descent rate is over 1 metre per second. The spacecraft is descending as leisurely as someone walking down a flight of stairs. The surface is almost within reach, with individual rocks coming into view. The spacecraft's sensors are detecting the ground's proximity with great sensitivity. The descent thrusters will switch off when the altitude sensors signal about the landing.

Contact. The spacecraft's legs touch the ground with a soft thud. Thrusters hush. Engines surrender their power, leaving a serene silence in their wake. The cacophony of combustion is gone at last. The Chandrayaan-3 had made its landing.

What followed revolutionised our understanding of the Moon. The ChaSTE instrument conducted precise measurements of temperatures at the landing site. On the sunlit slope, the surface reached a toasty 70 degrees Celsius during the lunar day. Yet, 8 centimetres below, the temperature dropped to a chilly near-freezing. That's a temperature swing of nearly 80 degrees Celsius in less than 10 centimetres. This discovery revealed the moon's surface layer as a highly non-conductive blanket, where heat struggles to travel through the regolith. Such a steep thermal gradient further indicates that even minor variations in terrain can create microenvironments with large temperature differences. The implications were immediate: sloped areas exceeding 14 degrees, even those enjoying some sunlight, might maintain subsurface temperatures low enough to support stable water ice. Scientists had believed for a long time that ice existed solely in the shadows of craters, where temperatures plummet below -200 degrees Celsius. Thanks to Chandrayaan-3, we have discovered ice could thrive far beyond our previous imaginings. The Pragyan rover has discovered an alluring elemental composition, containing low sodium and potassium but enriched with sulfur. This finding represents the *first* direct measurement of sulfur at the lunar south pole, indicating that this site exposes primitive mantle material. The analysis of the crater shows that the Shiv Shakti Point is approximately 3.9 billion years old, placing it in the ancient Imbrian Period, close to the era when life *first* may have emerged on Earth.

Isn't it interesting to see how *Space* as a concept resonates across every culture? It is the limitless expanse that holds all existence stretching without borders through time. For centuries, scientists and philosophers have mused over its nature. *Consider* the Moon: for countless generations, it has been our nocturnal companion. A luminous orb, it dazzled and inspired our curiosity. We have calculated its distance and charted its phases and eclipses. Yet, the secrets of its composition still elude us. We could observe only from Earth, limited to one angle of view. But that barrier is beginning to crumble. We are on the brink of a new era, as our connection to space transforms in profound ways. To a child, space is pure wonder. Each point of light becomes a question shimmering in the dark. The infinite void between stars looms large, both breathtaking and fearsome. For an astronomer hunched over spectral data at 4 am, space is a treasure trove of information. A redshift in the hydrogen emission line unveils secrets like velocity, distance, and the universe's relentless expansion. To a space professional building spacecraft thermal systems, space is a chilling vacuum. It conducts no heat, while radiation gradually degrades electronics. Micrometeoroids zip by at dizzying speeds turning constraints into a rigid reality.

However, space, at its purest essence, is far more than any single viewpoint. It's a three-dimensional vast expanse where all matter exists and where events unfold. Mathematically, it is described as a continuum—a flexible fabric that warps in the presence of mass, as elucidated by Einstein's general relativity. Where mass exists, space contorts into curves. These curves dictate the path of other masses. This is gravity: not a force pulling you down, but the shape of spacetime cradling your every step.

But there's another aspect of that landing I'd like to highlight, one that is fundamental and usually goes unnoticed. Spacecraft orbits always hinge on a precise collection of six numbers. Three of these numbers reveal their exact location, showing where they are in the vastness of space. The other three numbers delineate their velocity, showing how fast they are moving in every direction. It's always six numbers: never less, never more. This principle holds firm across the cosmos, from Apollo's missions to rovers on Mars, satellites orbiting Earth, or even probes venturing into Jupiter's enigmatic mysteries. In the world of space travel, six parameters are essential to chart any path. Why six? The textbook answer is surprisingly simple: space has three dimensions. Three coordinates map our position; three velocity components navigate our path. Simple multiplication does the trick. The math flows in a seamless manner, with predictions aligning perfectly with observations. This clarity lets us venture into more fascinating questions ahead. But this explanation only shifts the mystery further: why is space constrained to three dimensions?

Throughout my educational journey hitherto, I rarely paused to ponder this question. Space exists, after all, in three dimensions and it is the canvas of our reality. Asking "why three dimensions?" feels like questioning the very essence of the universe's existence. Or why mathematics clings to its truths, or why there's something instead of nothing at all. Such thoughts ignite long intriguing late-night debates. But they often fall outside what physics can measure, a discipline grounded in the concrete and measurable actions. I confess I made a grave error. The *first* hint of doubt arose during my engineering studies when I came across the entropy formula for black holes. There was something unsettling about it. The formula indicated that the information content of a black hole scales with its surface area, rather than its volume, as one might expect. It was perplexing that two dimensions could encapsulate the entirety of a three-dimensional space. I still remember sitting in our McClay library, feeling iffy. I kept checking the calculations, convinced there was a hidden error lurking somewhere. The mathematics was telling me something I did not want to hear: that the three dimensions I took for granted might not be fundamental. I think that unease is where this book really began. How could three-dimensional space fit comfortably within a two-dimensional boundary? Honestly speaking, it felt like a seismic shift in everything I thought I knew about reality. The result consistently emerged across various contexts, including black holes of all types, cosmological horizons, and flat spacetime when viewed from a clever angle. The relationship between information and area was not unique to black holes, but also revealed fundamental truths about the nature of space itself. Since then, evidence has been steadily accumulating. Holographic duality, the mathematical connection between gravitational theories and quantum theories, has emerged as a pillar in the world of theoretical physics. Today, scientists are creating *toy universes* in laboratories around the globe. These miniature constructs allow physicists to observe geometry emerging from the alluring collaboration of quantum entanglement. At the same time, gravitational wave detectors are pushing boundaries, exploring whether spacetime itself may be built from the quantum patterns that these theories propose. As you read these words, we are in the middle of a silent revolution. It is not the flashy type with big announcements or instant Nobel Prize fame, but rather a deeper transformation.

Space itself, the backdrop we have taken for granted since Newton, might not be fundamental. It appears as the stage for physics, the geometric structure that Einstein rendered dynamic with general relativity. But what if it emerges from something deeper? Much like temperature, which arises from molecular motion, or say a diamond's strength, which comes from arranged carbon atoms, space may be wrought from deeper quantum foundations. What stirs beneath: entanglement patterns, information geometry, or something even stranger? What we are discussing is quantum information. Here's the gist: We can think of space, which exists in three dimensions, as a canvas of geometry. In this space, quantum entanglement settles in and finds its balance under the rules of realistic physical conditions. This surely isn't the handiwork of a cosmic architect who settled on the number three. Nor are we just looking into a single universe in an expansive multiverse. It's certainly not only because three is the magic number for complex life, although that point does quite resonate. Quantum information spirals into a three-dimensional geometric formation. And this isn't just a random twist of fate but an actual dynamic process. However, such a bold claim needs to be carefully examined.

Think of entanglement like two magic coins. You flip both and seal each in a shoe box without looking. You take one to Ahmedabad and your friend takes the other to Nagpur. When you open yours and find heads you know immediately without any signal that your friend will find tails. But here's the quantum twist: until you opened your box, neither coin was heads or tails. They were in a state where they could be both at the same time. The act of looking made your coin choose and at the same time, it made the distant coin choose the opposite. This is not because information travelled faster than light. It's because the two coins were never separate; they were one entangled system that spanned across space.

So, let me be exact. Space is real. Emergence does not mean illusion. Chapter 3 is precise about this distinction. It's as real and tangible as anything, emerging from molecular interactions. When you navigate through three-dimensional space, you're essentially engaging with something that has physical reality—an emergent structure, but no less actual for it. I won't claim we have solved quantum gravity; that enigma still hangs in the air. Unifying quantum mechanics and gravity remains one of the deepest challenges in physics. Yet we have gathered enough clues to see the outlines of our answer. I'm not asserting this is irrefutable; that's not how science works. What I mean is that the evidence hath reached a tipping point. Multiple independent approaches—black hole thermodynamics, holographic duality, quantum error correction, laboratory quantum simulations—all point towards the same conclusion: space is emergent. Each of these avenues tells a convincing story that deserves our attention.

Consider how the early aerospace engineers who built the attitude control systems for the *first* deep-space probes were successful in their efforts. Their instruments, including sun sensors and horizon sensors measured angles on flat planes and the resulting equations proved effective. The spacecraft performed as expected, and the mathematics underlying their design was complete and self-consistent. And yet, the engineers' understanding of Euler's equations of three-dimensional rotation, which describe the relationships between roll, pitch, and yaw, was incomplete. As spacecraft began to execute complex manoeuvres, the control laws they had established did not fail due to error, but rather due to their limited scope. The incorporation of gyroscopes ultimately revealed the third dimension that had been present all along, but was previously undetectable by their instruments. Here is what matters: these engineers were not flawed in their approach, but they were working within a framework that, although limited, was internally consistent and allowed them to achieve their goals. This raises a provocative question: could we be those engineers? Not because we are missing spatial dimensions that we could look up to see, but because the three dimensions we experience might themselves emerge from a lower-dimensional description, much like a hologram encoding 3D information on a 2D surface. For instance, we navigate three spatial dimensions. General relativity describes gravity as the curvature of spacetime. Quantum field theory treats particles as excitations in fields that permeate space. Both theories work extraordinarily. GPS satellites confirm the predictions of relativity and particle colliders validate quantum mechanics with exceptional precision.

Yet what if our three-dimensional space, like those early control equations, is not fundamental? Not because hidden extra dimensions that we cannot perceive embed us, but because something more basic, which has no intrinsic geometry at all, gives rise to dimensionality itself as an emergent property? Modern physics hints at this intriguing possibility. That something deeper appears to be quantum entanglement, the correlations between particles that Einstein famously dismissed as "spooky action at a distance," but which decades of experiments have confirmed as fundamental to quantum mechanics. The insight is that the structure of entanglement between quantum systems seems to determine the geometry of space itself. In certain theoretical models, particularly those emerging from string theory's *holographic principle*, you can calculate spatial distance directly from entanglement measures. Just as a bar of iron becomes a magnet when enough atomic spins align not because any single atom is magnetic, but because the collective alignment creates a field, geometry emerges from quantum correlations. Enough entanglement, structured in the right way and space appears. More entanglement between two regions means they are closer in the emergent space, while less entanglement means a greater distance. This is not mere speculation. For example, the *Ryu-Takayanagi formula* precisely relates *entanglement entropy* to geometric area in specific quantum systems. The mathematics is surprisingly elegant: the same equations that describe how entangled particles correlate also describe the curvature of spacetime in Einstein's relativity. So in this view, space is what quantum information looks like at macroscopic scales. And the implications are profound, which suggests that our entire conception of space may need to be rebuilt from the ground up.

Our journey will persuade you of three essential insights:

First, the dimensionality of space requires explanation. Most people, including many physicists, take three spatial dimensions for granted, believing them too basic to be questioned. However, this viewpoint is reversed. There is nothing in our current theories that dictates this. The number of dimensions is a fundamental aspect of reality that our physics should explain, rather than assume.

Second, quantum entanglement generates spatial geometry. This is no idle speculation, as it is based on detailed calculations in quantum gravity that have been tested in laboratory quantum simulators. When physicists create specific entanglement patterns in quantum systems spatial relationships emerge. What started as abstract correlations between quantum bits becomes a measurable distance. Under the right conditions, these correlations produce a three-dimensional geometry.

Third, these ideas can now be tested. Gravitational wave detectors are probing spacetime at sensitivities unimaginable a decade ago. Quantum simulators are creating emergent geometries in controlled laboratory settings, which allow researchers to manipulate entanglement patterns and see how spatial dimensions respond. Within the next decade, experiments will either confirm that space emerges from quantum information or completely rule it out. Either way, physics will advance.

Here's our road map for what follows next. We shall begin with the puzzle itself. Why does the dimensionality of space need explaining? What goes wrong in universes with different numbers of dimensions? I'll argue that three dimensions hold a unique position: two dimensions cannot support enough complexity for structures like atoms or brains, while four or more dimensions lack the stability needed for planetary orbits or structures that last. Three dimensions sit at a special boundary between these extremes. From there, we turn to Einstein's spacetime and its limitations. General relativity has been one of the great successes in physics, as it has been able to predict gravitational waves and the behaviour of black holes with a high degree of precision. Nevertheless, the theory treats spacetime as a fundamental entity, a fixed backdrop against which the principles of quantum mechanics are played out. The existence of black holes suggests that this understanding of spacetime may be incomplete. This gives rise to a question posed by the information paradox: does spacetime extend to the smallest scales, or does it cease to be valid before reaching the singularity? This leads us to the holographic principle, where three-dimensional physics is encoded in a two-dimensional description. And from holography, we arrive at the central argument of this book: entanglement. As read earlier, a greater degree of entanglement between regions corresponds to a smaller spatial separation, while less entanglement results in a greater distance. In this framework, distance is a measure of the structure of quantum correlations. The connection between mathematics, information theory, quantum mechanics, and geometry is unexpectedly strong. I will try to present the main ideas in a way that is easy to understand, without using too many equations, but still provides enough detail for you to follow my reasoning. We will then look at what is happening in laboratories right now. I will discuss what these experiments have revealed so far, and what they have not yet been able to tell us. We will also explore how scientists are using gravitational wave detectors, precision measurements from spacecraft, and observations from astrophysics to study the structure of spacetime. This could potentially reveal signs of quantum geometry. While the research is still ongoing, if successful, it would have a major impact on our understanding. On the other hand, if the results are negative, it would help to rule out some theoretical approaches. Finally, we will *consider* what it means for something to emerge from a fundamental level. If space can emerge from quantum entanglement, then what about time, matter, forces, and can they emerge too? Is it possible that we could one day create spacetime itself? And what would it mean for our understanding of our place in the universe if reality is something that emerges from simpler components?

At the time of writing, I am a first-year Aerospace engineering student at the Queen's University Belfast, and while writing this book I have tried to develop a unique perspective. As a non-theoretical physicist, I aimed to clarify the origins of these concepts, their development process and my contributions. I would like to extend my heartfelt gratitude to my favourite Sean Carroll, for galvanising me to pursue this daunting endeavour. The television series *Salvation* and *The Three-Body Problem* were two works of fiction that caught my attention and led me to these questions long before. They showed me that fundamental physics can be a rich source of inspiration for storytelling and the universe's inherent strangeness is more intriguing than anything an author might invent. Both works have a lasting impact on this book. I wanted to create a narrative that would engage curious readers, driven by the same sense of wonder and curiosity I felt when watching and reading those stories. Any remaining errors are mine.

Lastly, I will not pretend we have all the answers. There are still genuine uncertainties. The mathematics indicates certain directions, but it often falls short of reaching definitive conclusions. Some experiments have not been done yet. I will be clear about what has been established and what is speculative. When the evidence is strong, I will say so. When we are making guesses based on limited data, I will acknowledge it too. My main point is this: do three spatial dimensions come from the structure of quantum entanglement? Under the right conditions, quantum information naturally arranges itself into a three-dimensional geometry. If this is right, it changes our understanding of space. Space is a pattern that emerges: quantum correlations arrange themselves into geometric relationships. We do not move through space as separate entities; we are patterns within the same quantum structure that creates space itself.

This is going to be a quirky depiction, and I will share why it is factual. Let's proceed!

I

The Question

"Fair is foul, and foul is fair." — Macbeth

Two dimensions cannot hold an atom together.
Four dimensions cannot hold a planet in orbit.
Three is not a compromise. It is a constraint.

1

Six Numbers

Imagine giving directions. Not to your favourite restaurant, but to anywhere in the universe. How many numbers do you need? The answer is astonishingly minuscule: *three*. One number tells you how far north or south, another tells you east or west, and the third tells you up or down. *With just three numbers, you can literally locate any location in the cosmos.* The launch of astronauts Frank Borman, Jim Lovell, and Bill Anders on 21 December 1968 marked a watershed. They travelled towards lunar orbit in a compact Apollo 8 command module, which had a habitable volume roughly twice that of a Hyundai Santro, Hyundai's popular hatchback. This module carried them around 384,400 kilometres from Earth, the farthest humanity had ventured at that time. Back in Houston, Mission Control tracked them using six numbers: three to determine their position - latitude, longitude, and altitude above the lunar surface; and three to measure their velocity - north-south, east-west, and up-down movement. The choice of six can be explained easily: space has three dimensions, meaning a position requires three coordinates, and velocity adds another three, making it a basic multiplication problem. Using those six numbers and Newton's laws, they could predict the spacecraft's location at any future moment. The prediction was exact, not approximate. Only six numbers captured everything about its motion through space. We had always assumed the reason was simple: space has three dimensions. To fix the position of a vessel in the infinite dark: there is something both humbling and magnificent in that economy. The universe offers no excess coordinates, no redundant directions, no dimension beyond what is strictly required. It is as though reality itself had been edited to the bone. Yet the question "Why three?" reveals an enigma. The answer may be far stranger than our wildest imaginations. It is a truth so evident that it tends to go unnoticed, similar to the obvious facts that water is wet and clocks run forward. Throughout much of human history, we accepted three-dimensionality without question. Aristotle regarded it as self-evident. Euclid crafted his geometry atop this very foundation, offering no justification. Newton, revolutionising physics with his laws of motion, treated three-dimensional space as a constant, the fixed stage where the equations of motion would unfold.

Kaluza's Fourth Dimension

In 1919, the mathematician and physicist Theodor Kaluza sent Einstein a draft proposing: what if space had a fourth dimension? Kaluza was not suggesting a mystical fourth dimension. He meant a genuine physical dimension, but one that was so small we had never noticed it. To understand this, consider a pressurised hydraulic line that runs the length of a rocket to understand this concept. For the aerospace engineer calculating pressure waves along it, the line appears one-dimensional, with pressure travelling only forward or backward. However, the line has a circumference, which is a genuine second dimension, albeit circular. In theory, pressure waves can travel around this circumference, but in practice, they do not, because the tube is too narrow and exciting those circular modes requires more energy than the system typically encounters. The one-dimensional description is sufficiently accurate for all engineering purposes, and only a tiny probe, sensitive to extremely short wavelengths and small enough to fit inside the tube, can detect the presence of that hidden dimension.

What if our universe works the same way? We move through three large dimensions each day, and a tiny fourth dimension is curled up so tightly that even atoms do not register it. The mathematics are elegant. With four spatial dimensions, Kaluza unified Einstein's gravity and Maxwell's electromagnetism, a framework that required only one hidden extra dimension. The strong and weak nuclear forces would not be understood as fundamental forces until decades later. Einstein was hooked but remained cautious. The idea was beautiful, perhaps too beautiful. But if this fourth dimension existed, why was it so small? Why were there four dimensions instead of three or fourteen? Kaluza had no answer. His theory simply assumed the structure without explaining it. The question remained: why does space and time have their particular dimensionality? We are still asking.

Today, modern physics has dramatically revised our picture of reality. String theory, the leading attempt to unify quantum mechanics with gravity, suggests we live not in four spacetime dimensions but in ten spacetime dimensions (nine spatial plus one time) in the case of superstring theory, or eleven spacetime dimensions (ten spatial plus one time) in M-theory. Fundamental particles are not points but tiny vibrating strings, which require extra dimensions to oscillate within—six or seven hidden dimensions beyond the three we observe, each coiled up to be microscopically small. Loop quantum gravity offers a different perspective: it suggests that space itself might be quantised, comprising discrete units at the smallest scales, with three large dimensions emerging from something more fundamental. Some other theories propose that we exist on a three-dimensional membrane floating through higher-dimensional space, or that we are a hologram projected from a lower-dimensional boundary. The details may vary, but nearly every modern theory of fundamental physics suggests that three large spatial dimensions might not tell the whole story. Either hidden extra dimensions exist or something deeper gives rise to the dimensionality itself.

This brings us back to Apollo 8 and those six numbers. Mission Control worked with four dimensions: three dimensions of space and one of time, which together form "spacetime." Events do not occur just in space or at a moment but happen at specific points in this four-dimensional continuum. Even Einstein based his revolution on this: spacetime is fundamental. It curves and flexes, and responds to matter and energy, but it is always there.

Recent developments suggest even this might be incorrect. Spacetime itself, including all four dimensions, may not be fundamental. It could be something that emerges, an output of deeper physics, the way a hexagonal snowflake is not built from tiny hexagons but crystallises that way because of how water molecules minimise energy as they freeze. If spacetime is emergent, the question shifts. We no longer ask, "Why does the universe have three dimensions?" but instead, "What gives rise to three-dimensionality from the underlying physics?" Maybe three dimensions work the same way. Maybe three is what naturally emerges when the fundamental ingredients of reality whatever they are come together according to physical laws.

What Is So Special About Three?

To understand why, we *first* need to understand what is special about three. *Imagine* you are designing a universe. You control how many spatial dimensions to include. What happens if you choose two? In a universe with two dimensions, everything exists on a flat surface. The inhabitants can move north-south and east-west, but they have no concept of "up", their entire universe is a flat surface. Two-dimensional physics works to a certain extent. You can have particles, forces, and gravity. But as things get more complex, they start to break down. Take gravity. In our three-dimensional universe, gravity follows an inverse-square law: if you double the distance, the gravitational force becomes one-quarter of its original strength. If you triple the distance, it becomes one-ninth. The reason for this inverse-square law is geometry. Gravity spreads outward from a mass in all directions. In three-dimensional space, the surface area of a sphere increases in proportion to the square of its radius. So, when you are twice as far away, the gravitational influence spreads over four times the area. This means the force per unit area, or the field strength, decreases as one over the square of the radius. In two dimensions, things are different because spheres do not exist. Gravity spreads out in a circle. As the radius increases, the circumference grows linearly. If you double the distance, the gravity will cover twice the circumference. This means the force per unit length decreases as one over the radius, which is an inverse first power, not an inverse square. This changes everything. In our three-dimensional universe, planets orbit stars in stable ellipses. The orbit of Earth is nearly circular, and even comets that have eccentric orbits remain stable.

A gentle perturbation shifts a planet into a different ellipse with a slight variation. The orbits carry significant resilience. In two dimensions with inverse-first-power gravity, orbits exist, but they are fragile. Planets do not trace ellipses; they trace rosette patterns that precess around the star. Small perturbations cause large changes in orbits and can slowly spiral inwards. Two-dimensional solar systems would have a short lifespan. But it gets worse. The problem is not just about orbital mechanics; it is also about atomic physics. Atoms are stable in three dimensions due to the balance between the attractive nuclear forces and the repulsive electron-electron interactions. The Coulomb force, which is the electrostatic force between charged particles, follows an inverse-square law, like gravity. This balance enables electrons to settle into discrete energy levels around the nucleus, which form stable atoms. In two dimensions, the Coulomb force follows an inverse-first-power law, altering atomic structure. Hydrogen, the simplest atom, still exists, with a single electron able to orbit a proton. However, when more electrons are added, the situation deteriorates rapidly. The screening effect changes, and electron-electron repulsion overpowers the nucleus's pull in unforeseen ways, causing multi-electron atoms to become unstable or not to form at all. Without stable atoms beyond hydrogen and helium, you cannot have chemistry. This means there would be no molecules. If there are no molecules, there can be no complex structures. And without complex structures, there can be no life. Two dimensions fail not because they are inconvenient, but because they cannot support the basic elements that make up complexity.

What about four dimensions? Four spatial dimensions might seem like an Amazon Prime upgrade, offering more room, more degrees of freedom, and more possibilities. However, the universe does not work that way. In four dimensions, gravity follows an inverse-cube law. When the distance is doubled, the gravitational force becomes one-eighth as strong as it was originally. This is due to the surface of a four-dimensional sphere, a three-dimensional hyper surface, increases with the cube of the radius. As discussed earlier, this has catastrophic consequences for orbital dynamics. In three dimensions, orbits are stable, which means planets follow ellipses. These ellipses are stable under perturbations, so a small push will only slightly alter the orbit, and the planet will not spiral into the star or escape to infinity. In four dimensions, no such stability exists. Orbits are not closed and instead spiral either inward or outward. This means there are no stable planetary systems, no long-lived solar systems, and no gravitational patterns that last for billions of years. This is not a minor issue. It is a fundamental problem. Life needs a stable environment to exist. Stars have to burn for billions of years and planets need to follow consistent orbits. But in four dimensions, none of this is possible. The problem is deeper than just celestial orbits; it affects at the very foundations of atomic structure. In a four-dimensional space, the Coulomb force, the electrostatic attraction between a proton and an electron, obeys an inverse-cube law, which means the electron's potential energy decreases more rapidly, scaling as $-1/r2$. When this is applied to the Schrödinger equation, the result is catastrophic: the atom loses its stable ground state, and the electron's wave function collapses inward, falling towards the nucleus with no minimum energy to prevent its collapse. If the simplest atoms cannot exist, then chemistry and complex molecules are impossible, and without these, there can be no observers, which suggests that four-dimensional space is incompatible with the fundamental building blocks of matter. The combination of electron-electron repulsion and the nucleus's steeper potential creates an unstable balance. As a result, the nucleus either pulls electrons towards itself or scatters them away. This is unlike carbon, which has six electrons arranged in well-defined shells, and in this case, stable configurations do not exist.

Why Orbits Need Three

In 1873, Joseph Bertrand asked what seems like a narrow question: in how many force laws does a planet, perturbed slightly from its orbit, find its way back? Not spiral inward, not drift outward; actually return, and trace the same path again. His answer, arrived at through careful analysis of the equations, was two. Exactly two. The inverse-square law, gravity, electromagnetism and the force proportional to distance, what we now call Hooke's law. Those are the only options. Every other force law you can write down produces orbits that precess: the ellipse rotates, the path never closes, the planet never quite returns to where it started. Bertrand's theorem is not a curiosity. It is the *first* sign that something is doing selection on our universe. Because Bertrand's result is not a theorem about gravity. It is a theorem about three dimensions. Rewrite it in two dimensions, or four, and the set of stable-orbit force laws collapses to zero. There are none. No inverse-square gravity in four dimensions produces a stable ellipse. Three dimensions is not just one option among many. It is the only integer where any stable orbit is possible at all. In two, four, or other dimensions, Bertrand's theorem yields different results. None of these results offer the stable dynamics that we see in our universe. Having three dimensions is not just good enough, it is uniquely suited to support complexity. But there is something troubling about this picture. The instability of two and four dimensions appears to be a cosmic accident. We are lucky to have a universe with three dimensions, and it works. However, relying on luck is not enough for physics. When physicists encounter a parameter that requires fine-tuning, it usually means there is more to the physics than we currently understand. The need for fine-tuning suggests there is an unresolved issue, implying that some mechanism should naturally determine the parameter's correct value. *Consider* the mass of the Higgs boson, which is 125 GeV. Why is it 125 GeV and not 10^{19} GeV, the Planck scale at which quantum gravity becomes important? The fact that the mass is lighter than expected suggests there are delicate cancellations, hinting at some hidden physics. The full answer still remains unknown. Or take the cosmological constant, whose observed value is roughly 120 orders of magnitude smaller than the naive quantum field theory prediction. This is baffling because it implies we are missing something fundamental about vacuum energy. Dimensionality raises similar questions. Why do we have three dimensions, and not two or four? The anthropic principle provides a weak answer: it says we could not exist if it were otherwise. But this is not a dynamical explanation; it is just a selection effect. Unless... Suppose dimensionality is not something we can set in advance. Perhaps it does not need fine-tuning. Is it possible that three dimensions simply emerge from the basic laws of physics? This is a possibility that we are exploring. *Imagine* the fundamental laws are not limited to a specific number of dimensions. They can be expressed in simpler terms: quantum information, patterns of correlation, and degrees of freedom. Dimensionality emerges from the organisation of information. The question of why we have three dimensions has a similar structure to asking why iron becomes magnetic when cooled below a certain temperature. Above 770 degrees Celsius, iron's magnetic domains point in random directions and cancel each other out, leaving the material with no preferred direction. However, below the Curie point, something changes: the domains spontaneously align, and a single preferred direction emerges from millions of atomic spins that previously had no coordination. No external force imposes this direction. The magnet effectively chooses it, as the aligned state has lower energy. Similarly, three-dimensionality may be favoured because it is the state that quantum information settles into when energy is minimised and the constraints of quantum mechanics are satisfied, rather than being a deliberate design choice. Maybe three-dimensionality is where quantum information naturally settles. The constraints of quantum mechanics and information theory suggest that three dimensions are what work. If so, universes with two or four dimensions would not just be unstable; they would be highly improbable. This is because the fundamental laws allow information to crystallise only in a three-dimensional form, not because three dimensions are preferred. This can be tested. If dimensionality emerges, we should be able to:

1. Frame the fundamental theory without making assumptions about dimensionality.
2. Show how three-dimensional geometry emerges naturally from that theory.
3. Explain why two or four dimensions are not produced.
4. Make predictions that are testable in laboratories or through observations.

This framework is currently being developed in modern physics. The three dimensions we see in Apollo 8's trajectory fall into a Goldilocks zone. They are not too simple, like two dimensions, where physics works but complexity is limited. They are not too chaotic, like four dimensions, where nothing stable can form. They are just right: complex enough to have a rich structure, and stable enough for this structure to last. This is not trivial. Every star, every planet, and every atom depends on three-dimensional space. This geometry makes stable orbits and stable atoms possible. The universe's hierarchy relies on three-dimensional geometry. It spans from quarks to atoms, atoms to molecules, molecules to cells, and cells to organisms. This continues up to ecosystems, planets, solar systems, and galaxies. If you change the dimensionality, none of it works.

When I say "two and four dimensions fail," I am not just talking about things like orbiting planets or unstable atoms. I mean that geometry cannot form in two or four dimensions under realistic conditions. Such universes are not only uninhabitable, they are also highly improbable, or at least highly improbable.

So why three dimensions? One answer is anthropic: we observe three dimensions because they are the only ones that allow observers to exist. A universe with two or four dimensions would be too unstable or too simple to support complex structures. This means we would have no chemistry, no life, and no physicists to wonder about dimensionality. This reasoning is valid but unsatisfying. Anthropic arguments explain why we observe three dimensions—our existence requires it. But they do not answer the deeper question: why does the universe have three dimensions in the *first* place?

Imagine a lottery where you hold the only winning ticket. You can explain how you won: you bought a ticket and your numbers matched. But this does not explain why the lottery exists or why someone drew those particular numbers. Anthropic arguments work in a similar way. Our existence requires three dimensions. But what gives the universe this dimensionality? Why is the universe not a lifeless void with the wrong number of spatial dimensions? Some physicists propose a multiverse with varying dimensionalities. In this vast ensemble, every possibility exists somewhere. We inhabit a three-dimensional universe because it is the only type that can support observers like us. But this raises new questions: why does the multiverse explore different dimensionalities? What determines this distribution? And, at a deeper level, is dimensionality something that is preset, or does it emerge from the fundamental physics?

That is the question this book addresses. Recent developments in theoretical physics, along with laboratory experiments and precise measurements from spacecraft and telescopes, suggest something remarkable: space does not just have three dimensions—space is three dimensions because that is what emerges naturally. Dimensionality is not just a setting for physics. Space is not a setting for physics. It is a phase, a collective state of quantum information, as distinct from its microscopic constituents as ice is from the water molecules that compose it. Three dimensions occur naturally when reality's fundamental ingredients organise themselves in line with physical laws. What are the basic components of everything? They are not particles, fields, or even spacetime. Essentially, reality appears to consist of quantum information patterns of correlation between quantum degrees of freedom. When quantum information combines in specific ways, it leads to the formation of three-dimensional geometry.

When Other Numbers Fail

In the year 1884, Victorian schoolmaster Edwin Abbott published Flatland: A Romance of Many Dimensions. The protagonist is a square, a two-dimensional being that exists on a flat plane. He interacts with other shapes, such as triangles, hexagons, and circles. In his world, a third dimension does not exist. The concepts of "up" and "down" are meaningless. One day, a sphere from three-dimensional space visits Flatland. The square sees something impossible: a circle appears out of nowhere, expands, then shrinks and disappears. What he's seeing is a three-dimensional sphere passing through his two-dimensional plane. Then it starts as a point of contact, grows into larger circles as the sphere moves deeper and finally shrinks as it moves away. To the square, this appears to be magic. From our three-dimensional perspective, however, it's simple. The square cannot perceive the third dimension, which explains everything. Abbott wrote Flatland as social satire, critiquing the Victorian class structure. But physicists interpreted it differently. If we are like the square, embedded in three dimensions, could there be a fourth dimension that we cannot perceive? What physics would govern this dimension? Why are we not like the Flatlanders? Why is it necessary to have three dimensions? Let's start with what "*dimension*" means. A dimension is a direction that enables independent movement rather than being a mere combination of other directions. For example, on a flat surface, you can move north-south and east-west. These are two independent directions. Although diagonal motion combines both, it is not a separate direction. This means the surface has two dimensions. In the space around you, *consider* the directions up-down, north-south, and east-west. These are three independent directions. When you move diagonally upward, you combine all three. But is there a fourth direction not up, not forward, not sideways, but something entirely different? Mathematically, the answer is yes. Mathematicians routinely work in four, ten, or even a million dimensions. The maths don't care; you just add more numbers to specify a position. For two dimensions, you use (x, y). For three dimensions, you use (x, y, z). For four dimensions, you use (x, y, z, w). When you add the w-coordinate, everything works out. But physics is not just mathematics; it is also constrained by the way nature works and seems like nature appears to be selective when it comes to dimensionality.

To get things concrete: let's build a two-dimensional universe and see where it fails. *Consider* a perfectly flat plane, an infinite sheet with no thickness. Fill it with point masses that move freely and are attracted to each other by gravity. In two dimensions, gravity works differently. As we saw earlier, the gravitational force depends on how gravity spreads out from a mass. In three dimensions, gravity spreads out over the surface of a sphere. The surface area of a sphere grows as the radius squared, so gravity decreases as one over the radius squared (the inverse-square law). In two dimensions, gravity spreads out in circles. The circumference grows linearly with the radius. The force decreases as one over the radius to the inverse first power, not the inverse square. This change has a fundamental impact on orbital mechanics. In three dimensions, Newton solved the two-body problem in 1687 for the case where one body greatly dominates in mass (like the Sun and a planet). The mathematical centre-of-mass reduction using the concept of reduced mass was later consolidated by Euler and Lagrange, though the essential physics was already present in Newton's Principia. One mass orbits another, following an elliptical path, or a circle, parabola, or hyperbola in special cases. The equations are exact, which means Kepler's laws emerge naturally. Orbits are stable: if you perturb a planet slightly, it will shift to a new ellipse and still maintain periodic motion. Whereas in two dimensions, the story is different. We start with a star at the origin. Here, a planet at a distance r experiences a gravitational force given by $F = GMm/r$, where G is the gravitational constant, M is the mass of the star, and m is the mass of the planet. Angular momentum is conserved: $L = mr^2(d\theta/dt)$, where θ is the orbital angle.

If you solve the differential equations, skipping the calculus, you find that orbits exist but they are not ellipses. They are open curves that spiral around the star in rosette patterns, similar to a spirograph: loops that precess and never close. Circular orbits can exist in two dimensions, but the underlying physics is unstable in a deeper sense. The logarithmic potential permits bound states with arbitrarily large negative energy, meaning configurations exist where particles spiral inward without limit, releasing infinite energy, a sign that the two-dimensional physics is fundamentally pathological, even if a planet in a circular orbit does not immediately plunge inward.

However, they are fragile. Adding a second planet, which creates the three-body problem, makes the dynamics chaotic. Small perturbations cause large changes, and multi-planet solar systems in two dimensions would be unstable. There's a deeper issue. The inverse first-power force permits bound states to have arbitrarily large negative energy. Configurations can be constructed where particles spiral inwards without end, giving out infinite energy. This makes the physics pathological. However, the real limitation is that it is complex. Now, *consider* a two-dimensional creature that's trying to evolve a digestive system. You need a mouth for taking in food, processing it, and getting rid of waste. In three dimensions, a mouth is at one end, an anus is at the other, and they're connected by a digestive tract that runs through the body. In two dimensions, this just doesn't work. If you had a digestive tract, it would cut your body in two; that is, the top half would be separate from the bottom half, and you'd fall apart. In two dimensions, life is very limited. You might have a mouth, but not a body to connect it to, or a connected body with no mouth. You would not have a digestive system, a circulatory system, or any complex organs. Life in two dimensions is too restrictive. Now *consider* chemistry. Atoms bond in complex geometries in three dimensions. Carbon forms tetrahedral structures, with one atom at the centre bonded to four others, arranged like a pyramid. This geometry underlies organic chemistry, DNA, and life itself. In two dimensions, atoms are confined to a plane. This means chemical bonds can only form two-dimensional networks. The rich complexity of organic molecules, such as the folding of proteins and the double helix of DNA, cannot exist. You might be able to form simple molecules, but you cannot achieve the hierarchical complexity that enables life. Even atoms themselves struggle in two dimensions. You can solve the Schrödinger equation, which forms the foundation of quantum mechanics, in two dimensions. This gives you hydrogen-like atoms, where an electron orbits a nucleus, held together by an electric force, not gravity. The maths involved are similar to orbital mechanics, but with an inverse first-power force in a two-dimensional space. The quantum mechanical solutions are peculiar. Energy levels differ from what we expect in three dimensions.

The ground state, which is the lowest energy level, exists, so atoms do not spontaneously collapse. But excited states behave oddly. Perturbations affect them in unpredictable ways. The quantum stability that is essential for atomic physics in three dimensions is compromised. Many-electron atoms fail to function properly in two dimensions. The shell structure that organises the periodic table (the foundation of chemistry) depends on three-dimensional quantum mechanics. In two dimensions, electron shells reorganise differently, resulting in physics that does not support the familiar chemistry we know. Two dimensions are too restrictive. Physics works, but complexity is limited. There are no stable many-body dynamics. Topology cannot sustain complex structures. Three dimensions seem ideal. But what about four? Four-dimensional space is difficult to visualise. We are trapped in three-dimensional bodies and minds. The phrase "perpendicular to all three spatial dimensions" makes linguistic sense but defies visualisation. Mathematics handles it easily. In four dimensions, the position requires four coordinates: (x, y, z, w). Vectors have four components. Rotations occur in planes rather than around axes, and multiple independent rotation planes can exist at the same time. This is already strange. In three dimensions, rotations happen around an axis. A spinning top rotates around a vertical line. The right-hand rule applies: you point your thumb along the axis and your fingers curl in the direction of rotation. In two dimensions, you rotate clockwise or counter-clockwise, which is rotation around an axis perpendicular to the plane. In four dimensions, you can rotate in the xy-plane and, at the same time, rotate in the zw-plane. These rotations are independent; an object can spin in two completely different ways at once. The planes of rotation must be orthogonal to each other for the rotations to be truly independent and to commute, meaning they have no effect on each other. Our intuition about rotations completely breaks down. Set intuition aside. How do motion and gravity work in four dimensions? Gravity obeys an inverse-cube law. In four dimensions, gravity spreads over the three-dimensional "surface" of a four-dimensional sphere. The "volume" of this surface, which is a three-dimensional measure, grows as the radius cubed. The force decreases as one over the radius cubed. *Consider* orbital mechanics. We start with a star at the origin. A planet, at distance r, feels a gravitational force $F = GMm/r^3$. The planet has angular momentum, but this becomes a bivector in four dimensions - representing an oriented plane rather than a vector. Now, set up the equations of motion and solve. The result is disturbing: stable periodic orbits do not exist. There's one exception: a perfectly circular orbit at exactly the right speed. Gravitational and centrifugal forces balance exactly, which allows for indefinite orbital motion. But this stability is fragile. Even the slightest perturbation, such as increasing the speed by one part in a trillion, causes the orbit to fail. The planet will either escape to infinity or spiral into the star. No elliptical orbits exist. There are no stable multi-planet systems, only instability. The inverse-cube force law creates a potential energy well that is steep near the star and shallow far away, with no region where stable orbits exist. The problem extends to quantum mechanics. To illustrate this, solve the Schrödinger equation for a hydrogen atom in four dimensions, where an electron is bound to a proton by the same inverse-cube electric force. The quantum mechanics of the atom undergo a catastrophe, known as 'falling to the centre', in this scenario. Since the potential drops off as 1/r2, the atom has no stable ground state, leaving the electron without a minimum energy level. Consequently, the electron is pulled into the nucleus, rendering stable atoms and the chemistry of life physically impossible in four dimensions. The electron's energy is negative, which keeps it near the proton. However, these states are fragile. When a second electron is added, as in the case of helium, stability vanishes. In such cases, electrons either plunge into the nucleus or escape. No stable atoms, no chemistry, no complexity. The four-dimensional instability extends beyond gravity and atoms, and it is pervasive. Think of waves in three dimensions: they propagate cleanly. Sound, light, and water waves all radiate from their sources, weakening over time but keeping their shape. A sharp pulse travels clearly and stays crisp. In four dimensions, waves do not behave well. Unlike in three spatial dimensions, where Huygens' principle ensures that a sharp pulse propagates cleanly along the wavefront with no lingering tail, the wave equation in even spatial dimensions (including 2D and 4D) behaves differently. There, the retarded Green's function has support not only on the light cone but also inside it, causing a sharp initial pulse to develop a persistent decaying tail. As a result, signals in 2D or 4D universes smear out over time: sharp features distort, information garbles, and coherence is lost. Three spatial dimensions are thus the lowest where waves propagate sharply without such aftereffects. This means energy lingers behind the wavefront, causing a sharp pulse to develop a persistent tail and distort as it propagates. This leads to communication problems, as signals scramble. Maintaining coherence is also difficult. Or *consider* topology. In three dimensions, knots exist. Strings can tangle in ways that you cannot undo without cutting. Knot theory, a mathematical field, studies this complexity. In two dimensions, knots do not exist; there is insufficient room for curves to cross over each other without self-intersection. There is no way to create a structure that is genuinely knotted. In four dimensions, knots also don't exist, but the reason is different: there's too much space. You can untangle any knot by moving the string through the fourth dimension. To a Flatlander watching us, it would look like the string passes through itself. What's impossible in two dimensions becomes trivial in three. Knots might seem like mathematical curiosities, but they hold physical significance. These topological structures, which are shapes that cannot be smoothly transformed, provide stability. DNA coils and supercoils, and proteins fold into shapes governed by topological constraints. The behaviour of magnetic field lines is affected when they knot in plasmas. In four dimensions, topological stability breaks down. This means structures that should be locked in place can slip free, and it causes physics to become unstable. The most severe problem is atomic instability. In three dimensions, the Schrödinger equation for hydrogen is:

$$[-\text{hbar}^2/(2m)\ \nabla^2 - e^2/r]\ \psi = E\ \psi$$

The Laplacian operator ∇^2 has three spatial derivatives. The potential energy decreases as 1/r. The solutions to this are the familiar hydrogen wave functions, which include 1s, 2s, 2p, and 3d. The energy levels are given by -13.6 eV/n^2, where n is the principal quantum number. The ground state, where n=1, has an energy of -13.6 eV. For excited states, where n=2, 3, and so on, the energy climbs towards zero as n approaches infinity. These energy levels are stable, meaning an electron in the ground state will stay there unless it gains external energy. Electrons in excited states, on the other hand, emit photons and move to lower levels, but they do not fall into the nucleus. The Schrödinger equation takes a different form in four dimensions:

$$[-\text{hbar}^2/(2m)\ \nabla_4^2 - e^2/r^2]\ \psi = E\ \psi$$

The presence of four spatial derivatives in the Laplacian operator ∇42 has a substantial impact on atomic stability. As the electrostatic force drops off as 1/r3, the potential energy scales as −1/r2, leading to a critical instability. Quantum mechanics reveals that the kinetic energy required to confine an electron to a tiny radius r also scales as 1/r2, a direct consequence of the Heisenberg uncertainty principle. This identical scaling of kinetic and potential energies means they cannot reliably balance each other, resulting in an unstable ground state. When the attractive force is sufficiently strong, the electron will inevitably be pulled into the nucleus, with no minimum energy floor to prevent its collapse. Consequently, even the simplest atom, hydrogen, has no stable ground state in four-dimensional space. The kinetic and potential energies both scale as $1/r^2$ near the nucleus, so no minimum energy level exists physicists call this 'fall to the centre.' The electron is pulled inevitably into the nucleus, and the atom cannot persist. The absence of stable atoms, which are essential building blocks, renders the complex chemistry required for life physically impossible. The existence of four dimensions affects matter in a big way.

Here's a preview. Black hole thermodynamics reveals something startling: information about the interior of a region is encoded on its boundary. A three-dimensional volume can be completely described by the information on its two-dimensional surface, which is known as the holographic principle. The mathematics is precise, demonstrating that three-dimensional gravitational physics is equivalent to two-dimensional quantum theory without gravity and they are not just similar, but equivalent. This means we now have two descriptions of identical physics, where one of these dimensions is not fundamental, but emerges from the pattern of quantum information in the lower-dimensional theory. Let's extend this idea further. What if all three spatial dimensions emerge in this manner? Suppose there is no geometry to begin with; just quantum information existing in an abstract space. As this information gets organised, a three-dimensional geometry emerges. Physicists have constructed models that show this. When quantum information is arranged in specific patterns, a three-dimensional space emerges. The distances between points, the directions, and even the number of dimensions—all these emerge from the structure of the information. This reveals why we have three dimensions. It is not due to anthropic selection, but dynamics. Three dimensions emerge because they are the stable configuration. Two and four dimensions are not just unstable for planets and atoms, but geometry itself cannot settle into two or four dimensions under realistic conditions. Here's how it works. Quantum entanglement, which Einstein called "spooky action at a distance", determines geometry. When two regions have a high level of entanglement, they're considered "close" in emergent space. When the entanglement is weak, they are "far apart". This is rigorous mathematics in certain theories. The measure of correlation, known as entanglement entropy, has a direct relationship with geometric distance in emergent spacetime. What entanglement patterns create three-dimensional geometry? The answer comes from information theory, quantum error correction, and tensor networks. To put it briefly, three dimensions emerge naturally from realistic entanglement structures. Two dimensions are too simple, as they have insufficient capacity for complexity. Four dimensions are unstable because entanglement patterns cannot maintain a stable configuration. Three dimensions are the Goldilocks zone, and now we have a mechanism: entanglement structure determines dimensionality. For systems with the right properties, three dimensions emerge naturally. When I say "two and four dimensions fail," I don't just mean things like orbiting planets or unstable atoms. I mean that geometry itself cannot form in two or four dimensions under realistic conditions, at least, this is what the theoretical arguments suggest. Two-dimensional and four-dimensional universes are not just uninhabitable but they are impossible, or at least vanishingly improbable, though this remains a conjecture. Quantum information does not organise into those shapes; it crystallises into three-dimensional geometry because three dimensions are what works. In the following chapters, I will develop this argument. The three dimensions are not accidental. They are the configuration that inevitably emerges from the entanglement patterns of quantum systems according to physics.

As we come to the end of our first chapter, the unstable orbits in four dimensions are not just a coincidence. Are they a symptom of something deeper? That four-dimensional space struggles to form, this is a core problem. This struggle leads to orbital instability which is one of the consequences. Fundamentally, physics tends to produce three dimensions naturally. Reality has three dimensions. That is why Mission Control needed six numbers. Chapter 10 develops the argument for why. Three dimensions provide a stable configuration that is naturally selected by quantum mechanics and information theory. That's the story we're exploring. But before we dive deeper, let's look at Einstein's insights about space, time, and gravity and see where his theory reaches its limits, which guide us towards our ultimate horizon.

2

Anthropic Trap

Here's a mind-boggling thought: *the universe appears to be rigged.* Not rigged in the sense of a conspiracy or deliberate design. But rigged in the sense that its basic parameters seem to be absurdly well-tuned. If you change the strength of gravity by just a few percent, stars cannot form. Adjust the electron mass, and chemistry unravels. Shift the cosmological constant by almost any amount, and galaxies never coalesce. And then there is dimensionality. As we have seen, three spatial dimensions sit at a narrow sweet spot — and that raises an uncomfortable question. Is that just luck? Why? One answer is simple yet appealing: we can observe three dimensions because they are the only ones that make it possible for observers like us to exist. If the universe had a different number of dimensions, we would not be here to observe it. The reason we observe three dimensions is that our existence depends on them. This is the anthropic principle, which is logically sound. However, it might be completely missing the point. And the logic is as follows.

The Royal Flush Problem

In 1961, physicist Robert Dicke questioned why the universe appears to be roughly 10 billion years old (we now measure it at 13.8 billion years, but Dicke's point stands regardless of the exact figure). His answer was that stars take time to form. Heavy elements, like carbon, are forged in stellar cores through nuclear fusion. These elements are necessary for planets to come together. Life needs planets and carbon. Conscious beings need time to evolve. If the universe were only a million years old, there would be no stars, no planets, and no life. If it were a trillion years old, all stars would have exhausted their fuel. Either way, there would be no physicists to ask the question. A universe around 10 billion years old shouldn't surprise us, Dicke argued. This is not because 10 billion years is inherently special, but because it is the time scale required for observers to emerge. The age of the universe reflects our existence within it. This solves the problem of cosmic coincidences without introducing new physics. The fundamental constants in nature are calibrated just right for life to be possible. If they weren't, we wouldn't be here to wonder about it. Space has three dimensions, which is what makes it possible for life and observers to exist, as this number of dimensions provides the necessary complexity. The logic is reasonable, but it leaves a deep sense of dissatisfaction.

Let's explore what anthropic reasoning does. *Imagine* you are dealt a royal flush in poker: the ten, jack, queen, king, and ace, all of the same suit. The odds are about one in 649,740, which is an miraculous improbability. "*Why did you get such an improbable hand?*" someone asks. You could answer anthropically: "Someone had to be dealt a hand. Every possible hand is equally improbable; one in 649,740. I simply got lucky. If I had got a different hand, I would be asking about that one instead. There is nothing to explain." Technically correct. But if you get royal flushes night after night, this explanation does not hold up. You would suspect something a rigged deck, a dishonest dealer, or marked cards. And it keeps happening, despite being highly unlikely, suggests there is a mechanism at work that you are not aware of. In physics, "rigged deck" means a deeper structure, which can be a mechanism, a selection process, or a fundamental law producing the apparently improbable outcome. Anthropic reasoning explains why we observe what we observe; we exist, after all. Still, it does not explain why the universe is structured in a way that permits our existence in the *first* place. Apply this to dimensionality. The anthropic argument is that we observe three dimensions because only three dimensions support stable atoms, chemistry, and life. We would not exist to observe anything in other dimensionalities. This is why three dimensions need no further explanation. Valid. But incomplete. Why? The argument assumes the universe's dimensions can vary. It presumes the universe could have had a different dimensionality, but it happened to have three. Dimensionality is treated like an adjustable parameter, a dial that could have been set differently. If dimensionality is a free parameter, this reasoning works. Different universes could have different numbers of dimensions. We observe three dimensions because that is where observers can exist. This is straightforward. But what if dimensionality is not a choice? What if the laws of physics only permit three dimensions? Then anthropic reasoning becomes meaningless. It is like saying, "I observe that two plus three equals five because I can only do arithmetic in a universe where that is true." This is technically correct, but it misses the point. Two plus three equals five because it is logically necessary, not because of some cosmic coincidence. The central question: Is three-dimensionality an arbitrary parameter that anthropic reasoning can illuminate? Or is it a mathematical necessity requiring no anthropic explanation? The anthropic principle comes in two forms: a weak version and a strong version. Brandon Carter, a physicist, introduced the term 'anthropic principle' in 1974, building on the foundation laid by Dicke's earlier work. The weak anthropic principle states: "Our observations are conditioned on our existence." This means we can only observe a universe that is compatible with our survival. This is straightforward and widely accepted, even if it is somewhat redundant. The strong anthropic principle claims: "The universe must have properties allowing life to emerge because observers are necessary to give it meaning." This leads to philosophical and metaphysical questions—what does "must" imply, and for what purpose? Most physicists prefer the weak formulation, avoiding the strong version. The weak anthropic principle, however, can be applied in two distinctly different ways. One is observational selection. *Visualise* that a vast multiverse exists. In this multiverse, some universes have three dimensions, others have four, and some may even have seventeen.

Since observers can only exist in three-dimensional universes, it is not surprising that we find ourselves in a three-dimensional universe, despite its potential rarity. This is legitimate anthropic reasoning, if the multiverse exists and dimensionality varies across it. The other approach avoids deeper questions altogether. "Why does space have three dimensions?" "Because if it didn't, we wouldn't exist." This response puts an end to inquiry. It explains our existence rather than presenting it as a mystery that needs investigation. Let me show you the danger. *Imagine* you wake up to find a loaded gun pointed at your head. The hammer is cocked, the trigger is pulled, but by a stroke of extraordinary fortune, it misfires. You survive. "Why are you alive?" someone asks. The anthropic answer: "If the gun had fired, I'd be dead and couldn't have this conversation. Since I'm here, it means the gun must have misfired. Me being alive shows that." True, but that is substantially incomplete. You still need to ask why there was a gun, who put it there, who pulled the trigger, what caused the misfire, and whether you are safe now or if another threat is waiting. The anthropic answer explains your survival but does not explain the mechanism of danger. You do not know if you are genuinely safe or if there are other guns aimed at you. In physics, these deeper questions have weight. Discovering that the universe is finely tuned for life, with its constants calibrated and its dimensionality exquisitely balanced, invites the anthropic principle to explain why we observe these features. Yet, it does not explain the fine-tuning itself. Let's focus on dimensionality. We live in three dimensions. Anthropic reasoning states: "Of course you do; you couldn't observe anything in fewer or more dimensions." This is fair. But why does the universe have three dimensions to start with?

- Option A. Random cosmic accident, the dimensionality could have been anything; we have three just by chance.
- Option B. Multiverse selection, numerous universes have different dimensions; we are in a three-dimensional one because that's where observers can flourish.
- Option C. Physical necessity, the laws of physics make three dimensions

the only possibility. Asking "why not four?" is meaningless, and four was never viable.

Three Explanations, One Right Answer?

Anthropic reasoning is essential for Options A and B. It helps explain why we observe what we do, given that different dimensionalities are possible. However, it cannot distinguish between Options A and B, and it has no bearing on Option C. So, which is correct? Option A, a random accident, is highly unsatisfying to physicists because it treats dimensionality as a brute fact: "It simply exists." This halts inquiry, which contradicts four centuries of physics progress. When Newton revealed that the same force governs both falling apples and orbiting planets, he revealed a hidden unity. Maxwell unified electricity and magnetism. Einstein revealed that gravity is spacetime curvature, which is geometric, not a force that pulls through space. Physics seeks unification and a deeper understanding. Saying "it's random" is a last resort, not a starting point. Option B, multiverse selection, is more sophisticated, but it also brings substantial challenges. The concept of the multiverse has become increasingly popular over the past few decades, especially in the fields of cosmology and string theory. According to eternal inflation, our observable universe is just one of many bubbles in an infinite sea of realities. The string theory landscape consists of 10^{500} or more possible vacuum states, and each of these could potentially be a universe with its own unique physical properties. If the multiverse exists and dimensionality varies across it, anthropic reasoning becomes more convincing. We observe three dimensions not because three is special in itself, but because we live in a region where that is what exists, similar to finding ourselves on a planet with liquid water, rather than a planet that is a frozen ice ball or a scorched rock. However, the multiverse framework faces serious challenges.

First, is dimensionality really different across the multiverse? String theory tantalisingly suggests it might; different compactifications of extra dimensions could result in different numbers of large observable dimensions. But this idea still remains speculative. We have no empirical evidence for other universes, and we certainly do not know anything about their dimensional properties.

Second, the multiverse is perhaps too permissive. If anything can happen somewhere, it "explains" everything but explains nothing. Why are constants what they are? The answer is the multiverse. Why are there three dimensions? The answer is the multiverse. Why is there something rather than nothing? The answer is the multiverse. At some point, physics risks becoming metaphysics.

Third, the multiverse framework is unfalsifiable. This is because if we were able to observe other universes directly, they would have to be part of our universe by definition. A large enough multiverse can accommodate any prediction.

Fourth, the multiverse does not explain its own existence or structure. Why does it have universes with differing dimensions? What fundamental law causes this variation in dimensions? We have simply traded one mystery for another.

Option C of physical necessity is what I'm advocating. What if three dimensions are not arbitrary but necessary? What if dimensionality is determined by deeper physics, rather than being set as an initial condition? Maybe three-dimensionality operates similarly to other emergent phenomena. If reality's foundation is deeper than spacetime itself, in quantum information, patterns of entanglement, and abstract degrees of freedom, then dimensionality might emerge from how this information organises itself. The three dimensions are not a choice from a list, but the natural configuration the universe settles into. Having two or four dimensions wouldn't just be inhospitable but be configurations that the fundamental laws of physics do not permit. In this view, anthropic reasoning becomes irrelevant. We observe three dimensions not because our existence dictates it, but because that's how reality fundamentally operates. The question of "why not four?" has a mechanistic answer rooted in the underlying physics, the structure does not allow it.

Let me address a natural counterargument. You might say, If three dimensions emerge from quantum information, so what?

Why does quantum information have the properties that generate three-dimensional space? Couldn't quantum mechanics itself have been different? This is a fair challenge but as Chapter 1 showed, two and four dimensions are not merely less convenient than three; they are dynamically unstable. The question then shifts from "why three?" to "why would quantum mechanics ever settle anywhere else? Fairly subtle. Yes, you could take anthropic reasoning as far back as you like in the explanatory chain. But at some point, the explanation has to be based on something fundamental. If we can show that three-dimensionality is a result of quantum mechanics and information theory, which are the foundations of modern physics and have been proven by many experiments, we will have answered one of the oldest questions in physics. We have brought together different parts of reality under one common framework. *Imagine* this in contrast to anthropic reasoning and multiverse selection. These approaches try to explain dimensionality by referring to a hidden structure that is the multiverse, which is an ensemble of universes and a selection mechanism. If we can state that "three dimensions emerge from the structure of quantum entanglement given certain information-theoretic constraints," this constitutes a concrete, mechanistic explanation, even if it raises new questions about quantum mechanics itself. Physics always reaches a limit somewhere. Even a theory of everything would explain the universe in terms of certain basic principles. Those principles might still appear mysterious. The goal isn't to eliminate all mystery since that's impossible. The goal is to explain as much as possible with as few unexplained assumptions as possible. If we can ground dimensionality in quantum mechanics and information theory, it will be genuine progress. We have a better explanation for dimensionality, one that is based on more fundamental and well-tested structures, rather than just treating it as an arbitrary input. Anthropic reasoning does not accomplish this. It asserts that things are as they are because if they were not, we would not exist to observe them. This is not an explanation in the scientific sense but a logical observation about selection effects.

Here's another perspective on the anthropic trap. Suppose we can show that three-dimensionality emerges naturally from quantum information. This means we have strong theoretical and numerical evidence that realistic quantum systems with the right entanglement structure naturally favour three-dimensional geometry. The arguments we develop in Chapters 10 and 11 from stability analysis, error correction, and RG structure all converge on d=3, though these are not yet proofs in the strict mathematical sense. They are compelling heuristics and quantitative arguments. The claim is that three dimensions are the natural, stable, and most physically coherent configuration, not that alternatives are logically forbidden. In contrast, systems with other dimensionalities are unstable and cannot persist. Now someone asks: "Why do we observe three dimensions?" The anthropic answer is still available: "If it were different, we wouldn't exist to ask." But this answer has become empty. It is like saying, "I observe that two plus three equals five because if it didn't, mathematics would be logically inconsistent and I couldn't do arithmetic." This is technically true, but it misses the point entirely. Two plus three equals five by logical necessity. We don't need an anthropic explanation for observing this mathematical fact. The danger of anthropic reasoning is that it can be applied universally. Whatever we discover about the universe's properties, we can always add: "And we observe this because if it were different, we wouldn't exist." This makes anthropic reasoning impossible to disprove and therefore scientifically useless. A principle that can "explain" anything ultimately explains nothing.

To clarify, the anthropic principle does have legitimate uses. It is particularly effective in cosmology when used with caution. If we assume that eternal inflation and a vast multiverse exist, then we can work out the probability of different cosmological parameters and what most observers would typically see. Think about the cosmological constant: the energy density of empty space. It could have been vastly larger than what we observe. If the cosmological constant were orders of magnitude greater, the universe would expand so rapidly that galaxies could never form through gravitational collapse. This means there would be no galaxies, no stars, no planets, no chemistry, and no life. The observed cosmological constant is extraordinarily small, approximately 10^{-120} in natural units. This seems to represent extreme fine-tuning. But, if a multiverse exists where the cosmological constant varies across different regions, the situation changes. Most universes would have values that are incompatible with galaxy formation and thus life. We observe an anomalously small value because we necessarily inhabit a rare region where observers can exist. In 1987, Steven Weinberg used anthropic reasoning to predict a small but non-zero cosmological constant, a prediction made before any observational evidence was available. By 1998, astronomers had discovered the universe's accelerating expansion, providing direct evidence for what Weinberg had predicted: a small, positive cosmological constant. This discovery spectacularly vindicated Weinberg's anthropic argument, the cosmological constant is non-zero but small, falling within the range Weinberg's bound allowed for observers to exist. It was not a precise numerical prediction. Weinberg's argument established an upper bound of roughly a few hundred times the observed value but the fact that observation fell within this range, rather than being enormous as naively expected from quantum field theory, was a success for anthropic reasoning. This is an example of anthropic reasoning done correctly: it has a well-defined multiverse model, a clear probability distribution across possible universes, and it makes specific predictions that can be tested and either confirmed or rejected through observation.

But notice the requirements: for anthropic reasoning to be valid, it needs a rigorously defined multiverse model, a calculable probability distribution, and concrete predictions that can be distinguished from alternative explanations. Without these elements, anthropic reasoning turns into unfalsifiable speculation. And for dimensionality, we are not even close to this standard. We do not know if dimensionality varies across a multiverse. String theory suggests it might, through different compactification schemes, but this is still unconfirmed speculation. Even within string theory, we have not got a clear idea of the probability distribution over different dimensional configurations. We cannot calculate the likelihood that a randomly selected universe would have three large dimensions rather than two or four. We do not even know if such calculations are possible in principle. Anthropic reasoning about dimensionality currently makes no testable predictions whatsoever. It just says, "We observe three dimensions because if we didn't, we wouldn't exist to make the observation." This is a tautology masquerading as an explanation. It doesn't add anything. In contrast to this, deriving three-dimensionality from quantum information theory is an alternative approach. If this succeeds, we will have:

We need a concrete, mechanistic explanation for why three dimensions emerge from the fundamental structure of physics rather than being imposed as an initial condition. The theory should provide specific predictions for experimental signatures of emergent geometry that can be tested in laboratory quantum simulators, and possibly through precision measurements of spacetime structure. Deep connections between dimensionality and other features of quantum gravity, holography, and information theory provide a unified framework that explains apparently disparate phenomena. This framework gives a coherent picture of reality, where spacetime itself emerges from a more fundamental quantum information-theoretic substrate.

Anthropic reasoning offers none of this. It states: "Dimensionality is what it is. If it weren't, we wouldn't exist to ask." This brings scientific inquiry to an end. It treats our existence as an explanation rather than a phenomenon that needs to be understood in mechanistic terms. I'm not completely dismissing the anthropic principle. Observational selection effects are real and need to be taken into account. If a multiverse scenario exists, anthropic reasoning is a valid way to understand why we see what we see, rather than other possible alternatives that make sense. But reasoning about dimensionality using anthropic principles is premature. Before invoking selection effects across a hypothetical multiverse, we should *first* investigate whether dimensionality can vary. Perhaps having three dimensions is not a choice among many, but rather an inevitable consequence of deeper physics that permits no alternatives. This proposition can be tested. If quantum information naturally and inevitably leads to three-dimensional geometry under realistic conditions, we will have explained dimensionality without needing to invoke anthropic selection. Well, observers cannot exist in other dimensional configurations is a consequence of the physics, not the reason for what we observe.

A Lesson from Quantum Mechanics

Let's look at a historical analogy that would help gain clarity on this distinction. In the early 20th century, physicists faced a deep puzzle: why are atoms stable? According to classical electromagnetic theory, electrons orbiting atomic nuclei should keep radiating energy and spiral inward, which would cause atoms to collapse in a fraction of a second. However, atoms clearly persist indefinitely. What is the reason for this? An anthropic answer could be: "If atoms collapsed, there would be no chemistry, no molecules, no complex structures, no life, and no observers. This is why we observe stable atoms because without atomic stability, we wouldn't be here to observe them." That is perfectly true. However, from a scientific perspective, it is a dead end. It does not explain the mechanism of atomic stability. The genuine answer came from quantum mechanics. Niels Bohr showed that electrons occupy discrete energy levels rather than being in continuous orbits. The ground state, which is the lowest energy configuration, is stable because there is no lower energy level for the electron to move to. It cannot spiral into the nucleus because quantum mechanics does not allow for intermediate states between discrete levels. Erwin Schrödinger's equation provided a rigorous formalisation of this insight. Quantum mechanics revealed that atomic stability is a consequence of physical law, specifically the quantisation of energy levels and angular momentum, rather than a cosmic accident or anthropic selection. The anthropic explanation was always an option. It explained why we see stable atoms, considering we exist. But it did not say anything about how it works. Quantum mechanics, on the other hand, explained why atoms are stable and gave us many predictions to test. I view dimensionality in a similar way. The anthropic answer is always there: "We observe three dimensions because we couldn't exist otherwise." But this does not explain why the universe has a three-dimensional structure. The actual answer, if my approach is right, will be found in quantum information theory showing that the structure of quantum entanglement necessarily gives us three dimensions, rather than us just picking them arbitrarily. This means anthropic reasoning is more of a footnote than a full explanation.

Where does this leave us? The anthropic principle can function as a trap, ending our investigation prematurely. It's tempting to say "the universe is fine-tuned for life, or we wouldn't exist" and stop looking into why the universe has the specific properties it does. For dimensionality specifically: Weak anthropic reasoning tells us why we observe three dimensions, but it does not explain why space is three-dimensional at its core. Strong anthropic reasoning drifts into metaphysics, but it does not produce testable predictions. Multiverse anthropic reasoning could be powerful if we had a concrete and well-defined multiverse model with calculable probabilities, but we do not. Emergent dimensionality from quantum information theory may make anthropic selection irrelevant, as it suggests that three dimensions are the only stable configuration permitted by fundamental physics.

We are pursuing this final option. To reach it, we must *first* understand Einstein's conception of spacetime and identify precisely where his theory reaches its limits, pointing towards a deeper structure. This forms the basis of what follows.

Before we proceed, let me address one final objection. You might argue: "If three dimensions emerge from quantum information, is that not just anthropic reasoning in disguise? Quantum mechanics creates the conditions that allow observers to exist. If quantum mechanics were different, we might get four dimensions instead, and observers could not exist. We observe three dimensions because that is what quantum mechanics produces, and it allows us to exist." Yes, in principle, you can take anthropic reasoning as far back as you like. But at some point, the explanation has to be based on something fundamental. If we can show that three-dimensionality is a necessary result of quantum mechanics and information theory, the foundations of modern physics, which have been validated by countless experiments over decades, we will have answered one of physics' oldest questions. We will have explained dimensionality in terms of more fundamental principles that have a solid basis in empirical evidence.

Could quantum mechanics itself be different? Possibly, in a certain abstract metaphysical sense. Quantum mechanics, however, is tightly constrained by fundamental consistency requirements: unitarity, which conserves probability, locality, which prevents faster-than-light signalling, and the mathematical structure of Hilbert space. Modifying these principles does not yield an "alternative quantum mechanics"; instead, you encounter mathematical inconsistencies or theories that experimental evidence has already ruled out definitively. The universe must have a definite character at some level. The goal of physics is not to explain why reality exists instead of nothing; this may be unanswerable. Instead, the goal is to show how complex phenomena arise from simple, fundamental principles. If we can derive three-dimensionality from the structure of quantum information and entanglement, then we have what we set out to do: explain dimensionality in terms of more fundamental structures. Anthropic reasoning falls short in this regard. It simply states that things are the way they are because if they were not, we would not exist to observe them. This is not a mechanistic explanation, but rather a tautology related to selection effects. Now let's move beyond the anthropic trap. Let's explore this in greater depth: what is spacetime at its most fundamental level? Where does Einstein's theory originate, and at what point does it break down, revealing the need for something deeper?

3

The Argument Ahead

Let me directly state the central claim. This book argues that space has three dimensions because, under realistic physical constraints, quantum information organises itself into a stable equilibrium configuration that is a three-dimensional geometry. This is a strong claim. I believe it is correct and it is testable. Over the following chapters, I will develop this argument systematically. By the end, you will understand that dimensionality isn't a fixed feature of the universe but an emergent property arising from more fundamental principles grounded in twenty years of mathematical development. Physicists have made a discovery in certain well-studied theories: three-dimensional gravitational physics has been shown to be mathematically equivalent to two-dimensional quantum physics without gravity. The extra dimension emerges from the quantum information that exists on the boundary. The framework now aims to extend this idea. If one dimension can emerge, why can't all three? It explores whether geometry, which arises from entanglement in these theories, might also arise in our universe. The evidence suggests that the answer is yes. When you work through the mathematics, three dimensions could emerge naturally, and this is not by happenstance. If this is correct, it could mean we are on the verge of solving one of the deepest mysteries in physics: why space has the structure it does. But that's not all. If spacetime is derived from quantum information, then the things we think are fundamental, such as particles, forces, and the laws that govern them, might also emerge from something else. The Standard Model of particle physics, describing known particles and forces excluding gravity, might not be fundamental. This model could be an effective description emerging from the root structure of quantum information. The laws of physics, including energy conservation, momentum conservation and quantum mechanics, might derive from information-theoretic principles. This change marks a shift in our understanding of reality, but it does not replace our existing knowledge, as general relativity and quantum mechanics have a strong track record of success. Instead, it reinterprets the things that these theories describe. General relativity is about the geometry that emerges. Quantum field theory is about the particles that emerge in this geometry. The Standard Model is about the forces that emerge between these particles. Emergence reaches down to the level of quantum information, which may be the only fundamental truth. This goes beyond philosophical debate. The notion that spacetime emerges from quantum information makes testable predictions, and we are now testing them. In laboratories around the world, physicists build quantum systems, such as arrays of ultracold atoms, superconducting qubits, and trapped ions, where they can precisely control entanglement patterns. These are model simulations, far simpler than actual spacetime. But they exhibit a principle: geometry can emerge from quantum information. If you arrange qubits in specific entanglement patterns, distance appears and curvature emerges. In some cases, an effective gravitational force also arises. Scaling a system like this from 50 qubits to 10^{122} qubits, which is the estimated number of degrees of freedom in our observable universe, gives you actual spacetime, a three-dimensional reality embedded in four-dimensional spacetime. Meanwhile, gravitational wave detectors like LIGO probe the structure of spacetime. Each black hole merger and each detected signal tests whether spacetime behaves exactly as Einstein predicted or shows subtle deviations that hint at an underlying quantum structure. At the scales we can observe, spacetime appears to be smooth so far. But as we approach the necessary sensitivity, we may be able to detect quantum corrections. The next generation of detectors, including advanced instruments, space-based observatories, and sophisticated technology, could potentially capture spacetime exhibiting a grainy, discrete behaviour, revealing its emergent nature. Spacecraft act as probes to investigate whether space conforms to classical geometry or exhibits minute deviations that reveal its underlying structure. By tracking signals with precision, the Cassini probe, during its solar conjunction in 2002 while en route to Saturn, had collected data accurate to fractions of a microsecond. The technique of lunar laser ranging, which involves bouncing pulses off reflectors installed by Apollo astronauts, has enabled scientists to measure the Moon's distance with an accuracy of within a centimetre. Over the course of a year, Gravity Probe B observed the precession of gyroscopes within a satellite. The collective findings from these experiments corroborate the predictions of general relativity, with spacetime behaving as expected so far.

Where We Are Going

Here's a quick glimpse of our journey ahead.

Part II looks at Einstein's theory of gravity, general relativity, and its limits. Einstein found that spacetime is dynamic and curves when it responds to matter and energy. But he considered spacetime to be fundamental. Black hole physics, especially black hole thermodynamics, shows problems with this idea. Information seems to be encoded on surfaces, not in volumes, which means geometry may be simpler than it appears.

Part III forms the core of the book. I will explain the holographic principle, which is the idea that two-dimensional information can encode three-dimensional space. I will also show how quantum entanglement determines geometric distance. This will be based on rigorous mathematics in certain theories. Finally, I will argue that our three-dimensional world arises from realistic entanglement structures. This is where I'll present the central argument.

Part IV is about time. If space comes from quantum information, what about time? I will argue that time emerges in a similar way from how entanglement evolves and correlations spread. The "flow" of time means that quantum information becomes more entangled, more correlated, and more classical. This helps us understand why time has an arrow: we remember the past, not the future, entropy increases, and cause comes before effect.

Part V presents experimental evidence. Quantum simulators show emergent geometry. Gravitational wave detectors examine the structure of spacetime. Precision spacecraft tests provide additional constraints. What we're looking for, is quantum foam. The grainy texture of space at its smallest scales. Here's what we've discovered and what experiments are coming next.

Part VI explores the implications. If spacetime emerges, what else emerges too? Are particles and forces also emergent? Could the physical laws themselves be emergent? What does this mean for the future of physics? Can we engineer emergent spacetime by manipulating quantum information to create new geometries? What philosophical questions arise from this?

The pages that follow will take us from Einstein's spacetime and black hole thermodynamics, through concepts such as holography and quantum entanglement, to laboratory evidence and the philosophical implications that follow. So hopefully, by the end, you won't just know that space has three dimensions; you'll also understand the why. Be forewarned: the answer is stranger and more beautiful than either a cosmic happenstance or anthropic selection.

What I'm Not Saying

Let me delineate what I am contending and what I am not. I'm not saying spacetime is illusory. Three-dimensional space is real in the way that liquidity is real. No single water molecule is liquid. Liquidity is a collective property, it emerges when enough molecules interact in the right way. Emergence is not a retreat from rigour. When we say space emerges from quantum entanglement, we mean something precise and calculable, not a vague gesture toward complexity. The word carries the same weight it carries in thermodynamics, where temperature is no less real for arising from the chaotic motion of molecules. You can pour it, swim in it, drown in it. The fact that it arises from something more fundamental does not make it less real. It makes it explicable. Space is like this. Distance exists. Dimensions are concrete. However, these are emergent properties that arise from quantum information, rather than fundamental entities that are inscribed in reality from the beginning. The caveats from the Commencement remain in force: space is real even if it is emergent, quantum gravity remains unsolved, and nothing here is asserted as proven beyond doubt. What I am saying is that geometry can emerge from quantum information, and there is now a concrete testable framework that shows how. This is how physics works: you propose a framework, derive the consequences, and test the predictions. If the framework survives scrutiny and explains more than it assumes, then it gains acceptance.

II

The Cracks

General relativity treats spacetime as the stage on which physics unfolds.
Black holes suggest the stage is made of something deeper.

4

Geometry of Gravity

Einstein's greatest assumption was that spacetime is real. It is not real in the sense that it exists; space and time do exist in some sense, of course. It is real in the sense that it is fundamental. In general relativity, spacetime is foundational, a four-dimensional geometric structure that is a metric field curving in response to matter and energy. Everything else, including particles, forces and fields, exists within this spacetime. This assumption has been astonishingly successful for over a century, describing everything from the perihelion of Mercury, the tiny wobble in its orbit that Newtonian gravity could not explain, to the large-scale expansion of the universe, with every experiment since confirming the prediction to an extraordinary precision. However, it may be incorrect, and there is growing evidence to support this idea. The nature of spacetime may differ from our current understanding, emerging from a more fundamental source, much like sound arises from air molecules that individually do not produce sound, or a metal's conductivity emerges from electrons that, by themselves, have no awareness of being part of a wire. Sound and conductivity are not properties of individual components, but rather arise from their collective behaviour. Similarly, spacetime might not be an inherent part of reality's fabric at its deepest level, but instead, it could be assembled from something that, in itself, lacks geometry altogether. The same logic may apply to space. Three-dimensional geometry may not come from any fundamental "three-dimensionality" built into nature. It may arise only when vast numbers of quantum systems lock into a specific pattern of entanglement and a different pattern would give a different geometry, or none at all. The evidence that this is more than an analogy comes from an unexpected place: black holes. But before heading straight to the evidence, we must grasp Einstein's claim and the extent of its impact. What we are proposing is indeed radical: Einstein's seminal work, we argue, describes an emergent phenomenon. The theory's success matters here, because we cannot discredit it without a thorough understanding of its accomplishments and limitations. Every experimental success I will outline not only validates the notion that spacetime is real but also imposes a constraint on any alternative theory. To be viable, our proposed replacement must be able to replicate these experimental successes.

Gravity Is Geometry

To understand this evidence, let us look at what Einstein discovered: what general relativity says, how well it works, and where it breaks down. In November 1915, Einstein presented his general theory of relativity to the Prussian Academy of Sciences in Berlin, having spent eight years on false starts and breakthroughs to get the complete theory. Gravity, Einstein argued, isn't a force; it's geometry. Mass and energy do not pull on each other across empty space. They warp spacetime itself. Objects moving through this curved spacetime follow the straightest possible paths, known as geodesics, which we perceive as gravitational attraction. The Earth does not orbit the Sun due to an invisible force. Instead, it follows a geodesic through spacetime, which the Sun's mass has curved. The Earth is essentially in free fall, following its natural path through curved spacetime geometry. This was more than a new theory of gravity. The universe's physics is deeply influenced by space and time, which are dynamic and actively participate in its workings. Spacetime tells matter how to move; matter tells spacetime how to curve. And it worked. Within weeks, Einstein calculated Mercury's orbit using general relativity. Einstein gave us a universe in which the stage itself is a participant. Where mass does not move through space but sculpts it, and space in return commands the motion of mass. A conversation between geometry and matter that never falls silent. That is what general relativity is. And yet a theory so successful at describing the world may still be silent about what the world, at its foundation, is made of. For decades, astronomers were puzzled by Mercury's orbit, as it precesses slightly more than Newton's theory of gravity had predicted. However, Einstein's equations provided a perfect explanation, requiring no adjustments or additional factors, just a straightforward geometry. Four years later, Arthur Eddington's expedition observed a solar eclipse, which confirmed another of Einstein's predictions: starlight bends as it passes near the Sun because spacetime curvature deflects light. Einstein became an international celebrity. For over a century, general relativity has performed brilliantly. It accounts for planetary orbits, binary pulsar dynamics, gravitational lensing, black hole mergers, and the large-scale expansion of the cosmos. Yet general relativity rests on a foundational assumption: that spacetime is fundamental. We now suspect that this assumption may be wrong. To understand why, we need to look closely at what Einstein's theory says. It's not enough to just say "mass curves spacetime", as this is only a slogan. We need to understand how the theory works, what it predicts, and where it fails.

What Einstein Actually Said

Let's start with the basic concept: geometry. Before Einstein, space and time were considered separate entities. Space was the three-dimensional arena where things happened. Time was the universal clock that ticked uniformly for everyone (including aliens!?). There was a universal "now"—an absolute simultaneity that was shared across the universe. In 1905, Einstein's special relativity shattered this picture. The concept of a universal "now" disappeared. Observers moving relative to each other disagreed about which events were simultaneous. As a result, time became relative. But special relativity still treated space and time as somewhat separate, even as it mixed them. You could change between reference frames, but space stayed as space and time stayed as time. Then in 1908, Hermann Minkowski, who was Einstein's former mathematics professor, gained an insight proposing the idea of unifying space and time into a single four-dimensional continuum: *spacetime*. An event in spacetime is defined by four coordinates: three that are spatial and one that is temporal. You are not just located in space at a particular time; you exist at a specific point in spacetime. Your life traces a curve through spacetime, which is called a worldline. You are born at one event and die at another, with every moment in between represented along this curve. This isn't just abstract mathematics. Spacetime has a real geometric structure, with a metric that measures distances between events. In special relativity, this metric is fixed, being flat and uniform everywhere. This metric shows you how to work out the spacetime interval between two nearby events, which are separated by small distances dx, dy, dz in space, and a small time interval dt.

$$ds^2 = -c^2dt^2 + dx^2 + dy^2 + dz^2$$

The minus sign is what distinguishes spacetime from ordinary four-dimensional space. Time enters with the opposite sign to space, creating different types of intervals: positive (spacelike, with more space than time), negative (timelike, with more time than space), or zero (lightlike, the path that light rays follow). This metric defines the geometry of spacetime. In special relativity, it remains constant, being the same for all observers everywhere. Einstein's breakthrough involved making the metric dynamic. General relativity states that the metric is not fixed, and it varies from place to place. The spacetime itself can curve. What does spacetime curvature mean? Think of walking on Earth's surface as we all do, unless you are an alien. If you walk in a straight line, always forward and never turning, you trace a great circle. For example, walk along the equator from Ecuador to Kenya to Indonesia and back to Ecuador. You have circled the globe without ever turning. This path, which seems like a straight line, closes on itself because the Earth's surface is curved. Earth's surface is a two-dimensional surface that is curved within a three-dimensional space. The curvature can be visualised by seeing how the surface bends. Spacetime is different. It is not embedded in a higher-dimensional space that we can step outside to observe. Spacetime is everything. The curvature is intrinsic, detectable only through measurements made entirely within spacetime. How do you detect intrinsic curvature? You do this by comparing the behaviour of curved space with that of flat space. As we all know, in flat space, parallel lines remain parallel forever. In curved space, they do not. On Earth, for instance, if you start at the equator and head north along two meridians, they are parallel to begin with, both pointing north and separated by a constant distance. But as you continue walking, they converge. They eventually meet at the North Pole, showing that parallel lines can intersect. This is curvature, which is an intrinsic property, not a matter of bending into another dimension: parallel lines just can't stay parallel. General relativity describes spacetime curvature through the Riemann curvature tensor, a mathematical object that encodes how geodesics, the straightest paths in spacetime, diverge. In curved spacetime, particles that follow nearby geodesics, which are initially parallel, will accelerate towards or away from each other. This acceleration is due to gravity. Einstein needed an equation that related matter to spacetime curvature. He wanted to know how mass-energy determines geometry. Einstein sent four papers to the Prussian Academy of Sciences in four successive weeks, each one correcting the last. The equation that arrived at the end of that month,

$$G\mu\nu = (8\pi G/c^4)T\mu\nu$$

is actually compact enough to fit on a t-shirt and contains within it the bending of light around the sun, the precession of Mercury's orbit, the expansion of the universe, the existence of black holes, and the gravitational wave signal that LIGO detected a century later. No equation in physics has had a longer reach. Einstein then did something that physicists almost never do: he looked at what his equation predicted and decided he did not like it. The equation said the universe must be expanding or contracting. Einstein believed the universe was static. So he added a term the cosmological constant Λ to hold it still. When Hubble's observations confirmed the expansion in 1929, Einstein called the constant his greatest blunder. He was wrong about that too. The constant he invented to get the wrong answer turns out, at the right value, to describe the accelerating expansion we actually observe. The universe is expanding, and the expansion is speeding up, driven by something that behaves mathematically exactly like Einstein's blunder. On the left side, the Einstein tensor $G_{\mu\nu}$ represents spacetime curvature. This is constructed from the Riemann curvature tensor and the metric. Although the mathematics are based on differential geometry, the basic idea is simple: $G_{\mu\nu}$ shows how much spacetime is curved. Right side: the stress-energy tensor $T_{\mu\nu}$ describes matter and energy, including mass density, energy density, momentum density, and pressure, all in one mathematical expression. The equation says: curvature equals matter-energy. This means the curvature of spacetime is determined by the matter and energy that are present. But there's more. The equation is bidirectional. Curvature determines how matter moves. Test particles, which are objects small enough not to disturb spacetime, follow geodesics, the shortest paths through curved spacetime. Here's the cycle: Matter curves spacetime (the right side determines the left side). Curved spacetime guides matter's motion (the left side determines geodesics). Matter moves, changing mass-energy distribution. New distribution curves spacetime differently. Repeat. It is a self-consistent, dynamical system, where spacetime and matter evolve together, influencing each other. Such is the beauty of general relativity. The framework operates fantastically.

Mercury and the Moon

Mercury's orbit, Newton's gravity predicts that planets orbit in perfect ellipses if only the Sun's influence matters. But planetary orbits precess; the ellipse itself rotates slowly. Newton had predicted some precession due to the gravitational effects of other planets. The observations, though, showed that Mercury's orbit precesses 43 arcseconds per century more than Newton's predictions. Einstein calculated Mercury's orbit using general relativity. The curvature of spacetime near the Sun, which is stronger than Newton accounted for, produces exactly the observed anomalous precession of 43.0 arcseconds per century. The difference between the total observed precession of 574 arcseconds per century and the 531 arcseconds per century predicted by Newtonian gravity from planetary perturbations. No further adjustments were needed. Light deflection, starlight that passes near the Sun bends. Eddington's 1919 eclipse expedition measured this deflection and found it to be in agreement with Einstein's predictions, though the initial measurements had large error bars. Modern measurements using radio waves from quasars have confirmed the effect with high precision. Gravitational Time Dilation, clocks run slower in stronger gravitational fields because time itself has different flow rates, more on this in chapter 17. Neutron stars are extremely dense remnants of supernova explosions. Some of these stars become pulsars, spinning rapidly and emitting radio pulses like cosmic lighthouses. When a pulsar orbits a companion neutron star, the orbital period can be measured with extraordinary precision by timing the pulses it emits.

Russell Hulse and Joseph Taylor discovered this system in 1974. They observed that the orbital period was decreasing steadily, as the neutron stars were spiralling closer together. Why was this happening? General relativity predicts that when masses accelerate, they emit gravitational waves, which are ripples in spacetime. The orbit then shrinks as this energy radiates away. The observed orbital decay matched Einstein's predictions precisely. Hulse and Taylor were awarded the Nobel Prize in 1993. It was the *first* indirect evidence of gravitational waves. On 14 September 2015, the LIGO (Laser Interferometer Gravitational-Wave Observatory) detectors, which are giant laser interferometers in Louisiana and Washington, directly detected a gravitational wave for the *first* time. Two black holes, each about 30 solar masses, spiralled together and merged, 1.3 billion light-years away. The final black hole was slightly less massive than the sum of its progenitors, about three solar masses of energy having been radiated as gravitational waves. Those waves travelled 1.3 billion years across the cosmos. They reached Earth, causing space to compress and stretch by one part in 10^{-21} (one part in a billion trillion). The waves were detected by LIGO. The LIGO experiment tests for quantum corrections to spacetime at scales many orders of magnitude larger than the Planck scale, thereby constraining models of quantum gravity that violate Lorentz invariance, even though the Planck scale itself remains out of reach. The shape of the signal over time, known as the waveform, matched Einstein's predictions exactly. Since then, LIGO and its partner detectors, Virgo in Italy and KAGRA in Japan, have detected dozens of mergers of black holes and neutron stars. Each of these confirms general relativity. The theory works in the strong-field regime, where black holes are compact and spacetime curves dramatically. At the largest scales, general relativity describes the universe itself. The equations govern its expansion and contraction. In the 1920s, Edwin Hubble discovered that distant galaxies are receding. The universe is expanding, which is a prediction of general relativity when applied to cosmology. We can trace expansion backwards. If the universe is expanding now, it was denser and hotter in the past. Go back far enough and you will reach the Big Bang. General relativity describes the universe's evolution from that initial state to the present, predicting cosmic microwave background radiation, light element abundances, and the formation of large-scale structures. Every observation confirms the theory.

Where It Doesn't

Given this success, why do we question General Relativity? We are indeed not doubting the theory where researchers have tested it, as that is already established. What we are questioning is whether spacetime is really fundamental and whether it exists at its most basic level. General relativity makes these assumptions: Spacetime is a smooth continuum. There is no smallest length scale, so you can zoom in indefinitely and the metric will remain well-defined. It has no graininess or discrete structure. The metric is a real physical field and not just a mathematical convenience. The metric $g_{\mu\nu}$ is just as real as the electromagnetic field or any other field. Spacetime is four-dimensional. It has three spatial dimensions and one temporal dimension. This concept is built into the theory from the start. The metric is based on a 4×4 matrix, which is designed to handle four dimensions. Causality is defined by the metric. The causal structure, which determines the events that can influence others, is defined by light cones. The fundamental categories are timelike, spacelike, and null intervals. These assumptions work beautifully for describing gravity from millimetres to billions of light-years, yet they fail under extreme conditions as:

At the centre of a black hole lies a singularity, where spacetime curvature becomes infinite. The metric blows up, and the theory breaks down. This is not a calculation error, but rather a prediction of general relativity itself. If you follow a geodesic into the black hole, you will reach the singularity in finite time. The idea of infinite curvature is unphysical, so something must be wrong. If you trace the universe's expansion backward, you reach a moment 13.8 billion years ago when all of space was compressed to a point. The density, temperature, and curvature were all infinite. This is another singularity, another point where general relativity breaks down. General relativity is a classical theory that ignores quantum effects. However, at small scales, quantum mechanics takes over. The Planck length, which is roughly 10^{-35} metres, is where quantum fluctuations of the metric become vital. At this scale, spacetime should fluctuate, foam, and dissolve into quantum uncertainty. General relativity cannot handle this. It treats the metric as a smooth, classical field. When you try to apply quantum mechanics to gravity naively, you get nonsense infinities that cannot be removed, and predictions that depend on how you do the calculation. This is the quantum gravity problem. General relativity and quantum mechanics both work in their respective domains but clash under extreme conditions. Inside black holes, at the Big Bang, and at the Planck scale, both theories must apply but we lack a unified theory. Empty space is not empty; it has energy. Quantum field theory predicts that quantum fields fluctuate even in a vacuum. These fluctuations contribute to the vacuum energy density. If we calculate this vacuum energy, you get a number that is vastly larger than what is observed. This is the sharpest discrepancy between theory and observation in all of physics. The vacuum does have energy, expressed through the cosmological constant. This energy density is what drives cosmic expansion. However, the observed value is far smaller than the naive prediction. Why the mismatch? General relativity does not explain it. The theory treats the cosmological constant as an input, a number to be measured; but does not explain why it has the value it does.

These three failures, that are singularities, quantum gravity, and the cosmological constant, are not random. They follow a clear pattern. Each occurs when we take general relativity to its limits: infinite curvature, tiny scales, or the vacuum itself. If spacetime is the foundation of everything, these failures are puzzling. But if spacetime emerges from something more basic, we expect to see these failures. Any description that emerges from smaller parts will break down when we reach the scale of those parts, just like the idea of water as a smooth fluid breaks down at the molecular level. So, what are the basic building blocks of spacetime? The theory is not incorrect; it is just incomplete. General relativity provides an effective description in certain situations, but it fails in extreme cases. This is the pattern of physics. Newton's gravity reigned for centuries. Then, a small anomaly in Mercury's orbit pointed towards general relativity, a more complete theory that includes Newton's as a limit, in cases of weak fields and slow speeds. Now general relativity has revealed its own limits, including singularities, tension with quantum mechanics, and the cosmological constant problem.

But what could that more serious idea be? The search has been ongoing for decades. String theory, loop quantum gravity, and causal dynamical triangulations are different approaches with different starting points, mathematical frameworks, and philosophical assumptions. One insight has emerged from these approaches: spacetime is likely not a fundamental aspect. The metric, which is the geometric structure described by Einstein's equations, might be emergent. It could arise from something even more fundamental. At its deepest level, reality is not just spacetime with a metric, but something more. The quantum degrees of freedom and patterns of entanglement suggest that spacetime itself emerges as a large-scale, low-energy approximation.

A black hole represents general relativity at its most extreme. If you compress enough mass into a small enough region, gravity becomes so strong that even light cannot escape. The escape velocity exceeds the speed of light. Once you cross the event horizon, which is the black hole's boundary, return is impossible. Once considered exotic curiosities, black holes have now been confirmed by decades of observation. Stellar-mass black holes are formed from the collapse of massive stars. Supermassive black holes, with masses ranging from millions to billions of solar masses, are situated at the centres of galaxies, including our Milky Way. General relativity gives us complete descriptions. The Schwarzschild solution is used to describe spacetime around a non-rotating black hole. The Kerr solution, on the other hand, deals with rotating black holes. These solutions are exact, as they perfectly satisfy Einstein's equations. For a black hole with mass M, the event horizon is at the Schwarzschild radius: $r_s = 2GM/c^2$. For the Sun, this works out to about 3 kilometres. For Earth, the figure is about 9 millimetres. For a billion-solar-mass supermassive black hole, the event horizon is about 3 billion kilometres, roughly the distance from the Sun to Uranus. Cross the horizon and your fate is sealed. All possible paths through spacetime lead you inward. In a finite amount of time, as shown on your clock, you will reach the singularity at $r = 0$. At this point, spacetime curvature becomes infinite, and general relativity breaks down. You are then compressed to a density that is infinite. Or are you?

In the 1970s, Stephen Hawking made a startling discovery: black holes are not perfectly black. They emit radiation. The precise mechanism of this process is rooted in quantum field theory in curved spacetime, where the quantum vacuum behaves differently near the event horizon than it does far away, resulting in the horizon emitting thermal radiation. Although Hawking later cautioned that this description is not entirely accurate, it captures the fundamental concept that the horizon's geometry is responsible for generating particles. To an infalling observer, empty space is what they perceive, but a distant observer sees it as a thermal bath of particles, a result of the horizon geometry mixing positive and negative frequency modes through Bogoliubov transformations, which clearly demonstrate this effect.

The temperature of this Hawking radiation is inversely proportional to the mass of the black hole. Stellar-mass black holes have a tiny temperature, about a billionth of a degree above absolute zero. The temperatures of smaller black holes are higher. As a black hole radiates energy, it loses mass and shrinks, which causes it to heat up, a runaway process. The black hole eventually evaporates completely. Here's the paradox: throw a book into a black hole. The book contains information which is the text, its atomic structure, everything that makes it unique. This information is swallowed by the black hole. As the black hole evaporates via Hawking radiation, the radiation that is emitted is thermal, completely random, and characterised only by its temperature. Chapter 8 will discuss the Hubeny-Rangamani-Takayanagi formula and its quantum-corrected island variant, which track entanglement changes during evaporation. The key point to note here is that the holographic principle suggests the information that falls into a black hole must be recoverable at some point. No trace of the matter that formed the black hole, such as a book, remains. The information appears to be lost. But quantum mechanics says information can never be destroyed. I want to pause here on Arun Kumar Pati, because his name deserves more space than it usually gets in popular accounts of quantum gravity. Pati works at the Harish-Chandra Research Institute in Allahabad, now Prayagraj and in 2007, together with Samuel Braunstein, he proved what is now called the no-hiding theorem. The result is precisely this: if quantum information appears to vanish from a system, it has not been destroyed. It has not been hidden in correlations between the system and its environment, either. It must be entirely recoverable from the environment alone. Information does not disappear into the gaps. It moves wholesale from one place to another. The universe accounts for everything. When I first read the proof, it runs to a few pages, elegant and not technically overwhelming, what stayed with me was not the mathematics but the implication for black holes. If the no-hiding theorem is correct, and there is every reason to think it is, then Hawking radiation must carry the full information about whatever fell in. The information is not lost. It is just extremely scrambled. The difference between those two things is the difference between destroyed and hidden, and physics says they are not the same thing. This means that information is displaced, not lost. When applied to a black hole, this theorem implies that whatever falls in cannot simply vanish into correlations between the radiation and the hole, but the information must eventually surface in the outgoing Hawking radiation itself. The no-hiding theorem shows that the answer to how this happens must exist, even if it has not been found yet. The universe does not hide things, it only moves them to a different place. It can be scrambled, spread out, and made practically inaccessible, but it must be preserved. This principle, known as unitarity, says that quantum evolution is reversible. Black holes appear to contradict this. Information goes in, but it disappears. This creates the information paradox, which has puzzled physicists for decades. The resolution tells us something direct about the nature of spacetime itself. A telling clue emerged from a more careful examination of black hole thermodynamics. In 1973, Jacob Bekenstein proposed that a black hole's entropy, its information content, is proportional to its event horizon's area, not its volume. The area is what matters.

$$S = (k\_B\ c^3\ A) / (4\ G\ hbar)$$

where A is the horizon area and the constants fix the relationship. For a black hole with mass M, the horizon area is:

$$A = 16\pi\ G^2M^2 / c^4$$

So, entropy scales as M^2, meaning that when you add mass, the entropy increases with the square of that mass. If you wonder why area? Area scaling suggests something deeper: the information describing everything inside a three-dimensional region can be encoded on its two-dimensional boundary. *Imagine* that all the information about what fell into the black hole every particle, every detail is somehow inscribed on the event horizon. The three-dimensional interior is redundant. It's a hologram that is projected from the two-dimensional surface. This is the holographic principle, a radical suggestion that challenges our understanding of spatial dimensions.

General relativity treats spacetime as fundamental. The metric is a real field. We live in a four-dimensional space. Geometry goes all the way down. Yet, black hole thermodynamics indicates a different view. Information scaling with area rather than volume suggests that one dimension may be redundant, and that our three-dimensional reality could be represented by two-dimensional information. This may not be fundamental, spacetime is emergent if geometry emerges from lower-dimensional information. This is the crack in general relativity. It is not a failure of the theory, but a sign that it describes something emergent rather than fundamental. Einstein introduced us to curved spacetime. Hawking and Bekenstein exposed the underlying flaw. Now we need to understand what gives rise to spacetime. At this point, quantum mechanics enters specifically in the form of quantum entanglement. To start with, we must make the holographic principle more precise. This means we need to determine exactly how three-dimensional gravity relates to two-dimensional quantum physics.

5

Black Hole Revolution

Black holes are where general relativity falters in its own principles. General relativity, as we have seen, describes gravity with stunning precision. Nevertheless, the theory also predicts certain situations where it fails, such as singularities where the curvature is infinite and the equations no longer make sense. Every black hole has a singularity at its centre. General relativity confirms the existence of the singularity, but it does not tell us what happens there. This is a serious problem, and it is a fundamental limitation. Yet black holes reveal something more than just where general relativity fails. They show us not just that it fails, but why it fails, and what this tells us about the nature of spacetime. The insight comes from an unexpected place: thermodynamics. Black holes have temperature and entropy. Importantly, this entropy obeys the area scaling we encountered earlier, which hints that one spatial dimension might be redundant. What stuck with me was not just the formula, but the moment its implication sank in. I had seen the formula before, but it was the consequences that really struck me. If the walls of a room contain all the information about its interior, then the interior itself becomes superfluous in some fundamental way. This challenges our intuition about the purpose that the third dimension serves. In this chapter, we will explore what black holes teach us about spacetime. We will see that the interior of a black hole, which is three-dimensional, can be completely described by information on its surface, which is two-dimensional. If this holographic principle applies to black holes, then it might also apply to all of spacetime. To start with, let us find out what black holes really are, with the formation. *Consider* a star, a massive sphere of gas held together by gravity and powered by nuclear fusion. The fuel eventually runs out, at which point fusion stops, pressure fails, and gravity takes over. For stars with masses below 1.4 solar masses, electron degeneracy pressure halts their collapse. The star then becomes a white dwarf, which is an Earth-sized ember that cools over billions of years. For stars between 1.4 and roughly 2–3 solar masses, neutron degeneracy pressure stops collapse at a higher density. The star then becomes a neutron star, which is a sphere of neutrons with nuclear density, and it spins rapidly, sometimes emitting radio pulses as a pulsar. And for stars with masses over about 3 solar masses, the collapse cannot be stopped. Gravity cannot be resisted by any known force. The limit that separates survival from collapse was *first* calculated by a 19-year-old Indian student, Subrahmanyan Chandrasekhar, on a steamship travelling from Madras, now Chennai to Southampton across the Arabian Sea in 1930. Chandrasekhar discovered that a white dwarf can support itself only if its mass is below 1.4 times that of the Sun. If the mass exceeds this threshold, electron pressure fails, and nothing can resist gravity. The scientific community were dismissive of his findings. As we read, Arthur Eddington, a renowned astrophysicist of the time, termed the result absurd, claiming that nature would never allow it. However, Chandrasekhar was correct, and Eddington was wrong. It took the Nobel committee 53 years to acknowledge this. Every black hole discussed in this book, including their horizons, mergers, and entropy calculations, is built on the foundation laid by Chandrasekhar. Matter is compressed to densities that just keep getting higher. What happens then? In Newtonian gravity, a star collapses to a point of infinite density. However, general relativity introduces a decisive feature: the formation of an event horizon. The event horizon is not a physical surface, but a boundary in spacetime defined by causal structure. Inside this boundary, all paths lead inward towards the centre. Escape is impossible because to get out, you would have to exceed light speed, and also because there is no direction called "outward." All futures from here lead inward. Cross the horizon, and you seal your fate. General relativity predicts that you will reach the singularity, in a region where spacetime curvature becomes infinite and the theory breaks down, within a finite proper time. This is a black hole. The defining feature is not darkness, but the event horizon, which is the causal boundary separating the inside from the outside. Weeks after Einstein published his field equations in 1915, Karl Schwarzschild found an exact solution that describes spacetime around a spherically symmetric mass. This solution, known as the Schwarzschild solution, revealed something remarkable: if you compress enough mass into a small enough radius, you will create a region from which nothing, including light, can escape. In the case of a non-rotating, uncharged black hole, we have the solution:

$$ds^2 = -(1 - r\_s/r)c^2dt^2 + (1 - r\_s/r)^{-1}dr^2 + r^2d\Omega^2$$

where $r\_s = 2GM/c^2$ is the Schwarzschild radius and $d\Omega^2$ represents the angular coordinates. The horizon is at $r = r\_s$. At this radius, the metric coefficient $(1 - r\_s/r)$ vanishes, and its inverse becomes infinite. This indicates a singularity. This led to confusion for decades. The question was whether there was a physical problem at the horizon, or if spacetime itself broke down. No. The singularity at $r = r\_s$ is a coordinate artifact; it is just a result of how we label spacetime, not a physical problem. Change to different coordinates, like Eddington-Finkelstein or Kruskal-Szekeres, and the metric is smooth right across the horizon. An infalling observer doesn't experience anything special at the horizon - there's no infinite curvature, no barrier, just smooth spacetime. The actual singularity occurs at $r = 0$, where curvature diverges completely. At this point, general relativity breaks down and quantum gravity becomes necessary. Rotating black holes are described by the Kerr solution, which is more complex but has similar features: an event horizon, a ring-shaped central singularity, and the same basic causal structure. For decades, physicists viewed black holes as mathematical curiosities, likely to be unphysical. Nature, they thought, would surely prevent such an extreme collapse. Nature had other ideas. By the 1960s, astronomers had discovered quasars, extremely bright objects in distant galaxies. These objects are so bright they outshine entire galaxies, yet they occupy regions smaller than the solar system. The most plausible explanation for this phenomenon is that quasars are fuelled by accretion disks that surround supermassive black holes. When matter spirals into the black hole, it generates immense gravitational energy due to friction and compression within the disk before it crosses the horizon, resulting in the compact accretion region emitting far more light than the surrounding galaxy. X-ray observations have revealed binary systems in which a visible star orbits a massive invisible companion that cannot be a neutron star. This companion is a black hole, and it accretes gas from its stellar partner. Radio telescopes have traced jets of particles shooting from galactic centres at nearly light speed. These jets are powered by the rotational energy of spinning supermassive black holes. By the 2010s, these observations had become routine. Black holes are common, being the natural outcome of stellar evolution and galactic dynamics. However, the actual revolution was not about observation, but theory. The careful study of black holes showed something unexpected about spacetime itself.

The Temperature of Nothing

Hawking radiation, as Chapter 4 established, is thermal. It carries no information about what fell in at least not on the surface. That is exactly the problem.

$$T = \text{hbar } c^3/(8\pi G M k\_B).$$

As a black hole radiates, it sheds mass. The temperature increases as the mass decreases ($T \propto 1/M$), and this accelerates the radiation rate. This creates a runaway process: the black hole evaporates, heating up as it shrinks, and eventually explodes in a burst of radiation at a sufficiently small mass. Smaller black holes have higher temperatures and radiate more intensely. Primordial black holes, formed in the early universe with small masses, could be evaporating now. Whether we observe it or not, Hawking radiation follows from applying quantum field theory to curved spacetime on solid theoretical ground. Black holes do radiate; they have temperature, and they are thermodynamic objects. This raises a question: if black holes have temperature, do they have entropy? In thermodynamics, entropy measures disorder, which is the hidden information or the number of microstates corresponding to a given macrostate. In 1973, Jacob Bekenstein argued that black holes must have entropy. His reasoning was to *consider* what happens when you throw a hot cup of coffee into a black hole. The coffee has entropy, roughly 10^{25} in natural units (S/k_B), reflecting the enormous number of molecular configurations. When the coffee crosses the horizon, it seems to vanish from the universe, and an outside observer can no longer access this information. However, the second law of thermodynamics says total entropy cannot decrease. The disappearance of the coffee's entropy would, therefore, be a violation of the second law. Bekenstein's resolution states that the black hole itself has entropy, which increases when it swallows objects. The total entropy of the black hole and the universe remains constant or increases. What amount of entropy is present in a black hole? Bekenstein proposed it should be proportional to the horizon's area. The reason for this is that the horizon is the boundary hiding information. The amount of information hidden should scale with the size of the boundary. When Hawking calculated the temperature of black hole radiation, he could extract entropy from thermodynamic relations. In natural units, the formula is: $S = A/(4L\_P^2)$. The proportionality between entropy and area was proposed by Bekenstein, but the exact coefficient of 1/4 came only after Hawking's 1975 calculation of the black hole temperature $T = hbarc^3/(8\pi GMk\_B)$. By using the thermodynamic relation $dS = dM/T$, physicists could confirm the coefficient exactly. This combined result is the Bekenstein-Hawking formula: $S = A/(4L\_P^2)$. The entropy of black holes is proportional to their surface area. Here's why area matters to us. In ordinary thermodynamics, entropy is extensive since it scales with volume. The entropy of a box of gas is proportional to the size of the box. When the volume doubles, the number of molecules doubles, and so does the entropy. This makes sense. Entropy counts microstates, which are ways to arrange particles. More volume means more room, which means more configurations and more entropy. However, black holes are different. The area scaling holds. If you double a black hole's radius, which requires double the mass, the area increases by a factor of four, since area is proportional to the square of the radius. As a result, entropy increases by a factor of four, not eight, because it is related to area, not volume. This is something deep. It suggests that information about a black hole's interior—all the particles, fields, and quantum states that have fallen in—is encoded on the surface of the two-dimensional event horizon, not in the three-dimensional interior volume. The interior of a black hole is similar to a hologram that is projected from its horizon. The holographic principle states that the information in a region of space is limited by its surface area, rather than its volume. It is possible that one dimension is unnecessary and that three-dimensional reality can be represented by two-dimensional information. Hawking radiation creates a paradox. Thermal radiation is random, and all that can be known about it is its temperature. When a black hole emits this kind of radiation, it does not reveal what fell into it. Suppose you prepare a quantum system in a pure state of particles with definite quantum numbers, which are carefully prepared. You drop this into a black hole and wait for it to evaporate via Hawking radiation. You end up with thermal radiation that is a mixed state that is completely random. What was once pure has become mixed, and information has been lost. But quantum mechanics says this cannot happen. Quantum evolution is unitary, which means it is reversible and preserves information. A pure state will only evolve into another pure state, and never into a mixed state. The time evolution operator preserves norms, probabilities, and information. Yet black holes appear to violate this. Information goes in, but only thermal radiation comes out. Is information lost? This is known as the black hole information paradox. It has puzzled physicists for decades. The tension between the two theories seems irreconcilable. Hawking's calculation is solid. Quantum field theory in curved spacetime near a black hole produces thermal radiation, but no information escapes. Unitarity is fundamental, and every consistent quantum theory relies on it. If unitarity is abandoned, quantum mechanics collapses, and probabilities are no longer conserved, causing the framework to break. Things couldn't stay that way forever; something had to give. The resolution emerged over decades of work. In the 1990s, Gerard 't Hooft and Leonard Susskind proposed that information is not lost; it's encoded on the event horizon. When a particle falls into a black hole, its quantum state information gets imprinted on the horizon. As the black hole evaporates, this information escapes through subtle correlations in Hawking radiation. This seemed speculative until AdS/CFT, the gauge/gravity duality we will discuss in our next chapter, provided mathematical support. In this framework, black hole interiors and Hawking radiation are described as quantum states in a lower-dimensional theory, with information being explicitly conserved. This implies that information must be conserved in the gravitational description as well. A black hole's entropy increases first as it swallows matter, peaks around half the evaporation time, and then decreases as Hawking radiation carries the information out. The black hole eventually evaporates completely, leaving the radiation in a pure state; the information is fully recovered. Don Page worked this out in 1993, and the curve that bears his name is disarmingly simple. At *first*, as the black hole radiates, entropy grows; Hawking radiation comes out, it is thermal, it carries no information, and the system looks more and more random. But Page showed that if information is preserved, entropy must eventually turn around. Halfway through the evaporation, it should peak, then decline, until at the end the radiation is back in a pure state. The Page curve is what unitarity demands. For twenty-six years, no one could derive it from a theory of quantum gravity. The Hawking calculation gave the wrong answer entropy growing forever and nobody could find the mistake. Then in 2019, five physicists working partly independently, Ahmed Almheiri, Netta Engelhardt, Donald Marolf, Henry Maxfield, and Geoffrey Penington found it. The mistake was in not accounting for a subtle topological contribution to the path integral: a spacetime geometry called an island. When you include the island, the entropy turns around. You get the Page curve. The information is not lost. It was hiding in the topology of the path integral, in a geometry that Hawking's original calculation had not considered. The black hole information paradox, which had seemed intractable for half a century, was not a paradox. It was an incomplete calculation. They introduced the island formula, a quantum-corrected version of the Ryu-Takayanagi formula, which revealed that previously disconnected regions of the bulk spacetime, known as islands, start contributing to the boundary entanglement at late times. This momentous change leads to the entropy turning around and decreasing, precisely reproducing the Page curve that unitarity demands. Information does escape. Hawking radiation appears to be thermal only at early times. At later times, subtle quantum correlations carry all the information out. What matters here is that the black hole interior is not fundamental. The information about the interior is encoded on the horizon. This means the three-dimensional interior can be reconstructed from two-dimensional boundary data. This is the holographic principle in action. The interior of a black hole behaves like a hologram, being a three-dimensional image that is formed from the information on a two-dimensional surface. If this principle holds true for black holes, could it also hold true for all of spacetime? Leonard Susskind coined the term "black hole complementarity" to resolve the differences in perspectives. An observer who falls into a black hole does not experience anything out of the ordinary at the horizon. The spacetime here is smooth, so they can pass through it and eventually reach the singularity where it destroys them. This is what the experience is like from the perspective of the observer falling in. To an outside observer, the horizon reveals nothing crossing it. Matter seems to slow, become redshifted, and freeze at the horizon. From their viewpoint, nothing crosses over. Matter builds up at the horizon, becomes quantum-scrambled, and ultimately emerges as Hawking radiation. Both descriptions are valid—complementary, like wave-particle duality. You can't be inside and outside simultaneously, so no contradiction arises. They are different perspectives on the same physics, after all. The point is that the outside observer's description is complete. This includes everything that happens, such as the infalling observer's experience, which is encoded in the physics at the horizon. The degrees of freedom inside are redundant. This is holography: a three-dimensional space that is fully described by two-dimensional boundary data.

The *area law* for black hole entropy suggests a fundamental limit. *Consider* any region of space: what's the maximum entropy it can contain? Naively, it seems unlimited. Pack in more matter, compress it further, and entropy increases. But this approach fails. If you compress matter too much, it collapses to form a black hole. Once that happens, the entropy is fixed by the area of the horizon. Maximum entropy in any region occurs when it becomes a black hole. The formula for this is $S\_max = A/(4L\_P^2)$, where A is the boundary area. The holographic entropy bound, formulated by Raphael Bousso and others, states that the maximum information in a region obeys the same area scaling we saw for black holes. This is strange. In three dimensions, we would expect information capacity to increase with volume since the more the space, the more room for information. But the bound says information capacity is limited by surface area. This suggests that three-dimensional space might be redundant. If the fundamental description exists on a two-dimensional surface, then the three-dimensional interior emerges from it. Information capacity scales with surface area, not volume. The third dimension adds no new information. It just takes up room. General relativity treats them as classical objects, with smooth spacetime and deterministic evolution, and does not account for quantum effects, except for Hawking radiation, which arises from quantum fields in classical spacetime. Black hole thermodynamics, which includes entropy, temperature, and evaporation, requires quantum mechanics. The formula for entropy, $S = A/4$, involves hbar, or Planck's constant. The microstates have to come from a quantum theory of gravity. The singularity at $r = 0$ is also where general relativity fails because infinite curvature means the classical description breaks down, and so quantum gravity is necessary. String theory, loop quantum gravity, and other approaches try to explain black hole entropy by counting microstates. String theory has had successes, where it calculates Bekenstein-Hawking entropy for certain black holes by counting string states. The precision of this calculation reached its highest point with the work of Ashoke Sen, a renowned Indian string theorist at the Harish-Chandra Research Institute in Allahabad. Sen's development of the entropy function formalism enabled exact computations of black hole entropy, including quantum corrections that go beyond the leading area term for extremal black holes, which are poised on the brink of forming naked singularities. These corrections exhibit a distinctive logarithmic form. A separate calculation, based on counting microscopic string states, yields the same result, thereby validating the theory. This concurrence of two independent calculations, each using a distinct mathematical language, confirms that the theory holds in both mathematical languages. Sen's research has conclusively established that Bekenstein's entropy is more than just an analogy. It is an exact, quantum-mechanical concept. The fact that geometric and informational considerations converge on the same value for entropy is telling us something important that we have not yet fully grasped. Loop quantum gravity suggests that discrete microstates correspond to distinct quantum geometries, quantising the horizon area. At the Tata Institute in Mumbai, Spenta Wadia and Sumit Das were key physicists who built on the microscopic foundations of this programme in 1996. Their research showed that the characteristics of Hawking radiation from near-extremal black holes, such as temperature, spectrum, and rate, could be reproduced exactly by counting states of open strings on D-branes, objects with no surface similarity to a black hole horizon. This breakthrough demonstrated that two completely different approaches could yield identical results. These confirm that black hole entropy is real, physical, and quantum-mechanical. Black holes have microstates, and their thermodynamics are genuine, not just an analogy. The area law suggests that holography is a fundamental principle of quantum gravity. When physicists thought they had understood black hole information, a new puzzle came up in 2012. The "firewall paradox" was posed by Ahmed Almheiri, Donald Marolf, Joseph Polchinski, and James Sully. Hawking radiation is made up of entangled pairs, where one particle falls into the black hole and its partner escapes. The particles that escape are entangled with those that remain inside the black hole. But quantum mechanics puts a limit on entanglement. A particle can be maximally entangled with only one other system. If escaping Hawking radiation has entanglement with the black hole interior, as the thermal nature suggests, then it cannot be entangled with later radiation as well, which is what is required for information to escape through the Page curve. Something must give. The proposed resolution is that the event horizon is not a smooth boundary; it is a "firewall", a region of intense energy that destroys anything that crosses it. For an observer falling in, there is no smooth spacetime - the horizon incinerates them. This contradicts general relativity, which states that the horizon should be indistinguishable from empty space, as per the equivalence principle. The firewall paradox has sparked intense debate. The proposed resolutions are quantum error correction, ER=EPR (the conjecture that wormholes connect entangled particles), non-local effects in quantum gravity, and reconsidering our understanding of spacetime interiors. The debate continues, and a picture is emerging: black hole interiors aren't fundamental. They're approximate notions that break down when scrutinised carefully. The fundamental description exists at the boundary, where the interior emerges - reconstructed holographically from the boundary data.

What We Have Learned

To summarise what we have explored, let us go over the points. Black holes have entropy. This entropy follows the area law — proportional to the horizon's surface, as Bekenstein and Hawking showed. The entropy, S, is equal to the area, A, divided by 4 in Planck units. The preservation of information is highly probable. The black hole information paradox is generally considered to be resolved in principle, with the island formula and quantum extremal surface calculations providing strong evidence that information is released through Hawking radiation. Nevertheless, physicists continue to debate the precise physical mechanism, and a full proof of unitarity remains out of reach, despite its widespread acceptance among physicists. Information that falls into a black hole eventually escapes through Hawking radiation. This process works by encoding the information on the horizon, rather than storing it inside the three-dimensional interior. The interior is holographic. Everything about a black hole's interior can be described by physics on the two-dimensional horizon. This makes the three-dimensional interior redundant, as it is a holographic projection of the information on the boundary. Spacetime can be emergent. At least near black holes, the geometry of spacetime is woven from quantum correlations on a surface with fewer dimensions. The metric, curvature, and even dimensionality itself are derived, not fundamental. General relativity is incomplete. While it delightfully describes emergent geometry, it doesn't explain where that geometry comes from or what happens at singularities. A deeper theory, quantum gravity, is needed. These lessons are not just about black holes; they indicate something universal about spacetime. The holographic principle, which originated from black hole thermodynamics, might apply everywhere. Our universe has a cosmological horizon, which is a boundary beyond which we cannot see because space is expanding. Since this horizon has an area, it should have some entropy. Calculate the entropy with $S = A/4$ and you get approximately 10^{122}. This is the maximum entropy that our observable universe can contain. The observed entropy, which is dominated by black holes, is much smaller, at around 10^{104}. So, does this mean that our observable universe is a hologram? Is our three-dimensional space just a projection of the information on the cosmological horizon? Perhaps. The cosmological holographic principle is more speculative than its black hole version, as it lacks a tight mathematical formulation. However, the entropy bound is suggestive, implying that our three-dimensional universe could be encoded in two-dimensional information on a distant boundary. To be precise, we need mathematics. We require a concrete example where three-dimensional gravitational physics exactly matches two-dimensional quantum physics without gravity. We need to see holography in action in a controlled setting where we can calculate everything. That example exists, and it's called gauge/gravity duality, or AdS/CFT correspondence, which is the subject of the upcoming chapters. Emergent interiors, holographic encoding, and the primacy of boundary information lay the groundwork for the ideas that we will explore next.

6

Holographic Clue

In 1997, Juan Maldacena, a young physicist at Harvard, was working on a problem in string theory that seemed abstract. He wanted to understand how strings behaved in certain curved spacetimes. Maldacena's findings had a meaningful impact, changing everything. I intend to be meticulous about this moment: it transforms the question "why three dimensions?" from a topic of philosophical debate to a query that can be addressed with a clear answer. Maldacena's research found that two completely different theories: one describing gravity in five dimensions, another describing quantum fields in four dimensions, were secretly the same theory. They were not approximately the same, nor were they just similar. They were identical in every mathematical sense and could be proven as such. In the five-dimensional theory, each black hole corresponds to a specific quantum state in the four-dimensional theory. Each gravitational wave also has a counterpart in a quantum field. Calculations in one theory can be translated into the other with accuracy. The two theories are like different languages that describe the same physics. This is the AdS/CFT correspondence, and it's the closest thing we have to proof that spacetime itself emerges from quantum information. Technically, AdS/CFT remains a conjecture, it has never been rigorously proven in the mathematical sense. Rajesh Gopakumar, Director of ICTS in Bengaluru, has been systematically working to convert the duality from a conjecture to a derivation. His work shows that when the Feynman diagrams of a boundary gauge theory are reorganised, they get reassembled into the worldsheet of a string in a bulk spacetime, and the bulk is a natural outcome. If this programme is completed, it will establish that quantum information is necessary for spacetime to emerge, and not just a factor that allows it to happen. However, it has passed thousands of non-trivial consistency checks, and no physicist working in the field seriously doubts its validity. Here's why it matters: one of those five spatial dimensions does not exist in the four-dimensional quantum theory. It is not hidden or curled up; it simply does not exist. This dimension emerges from the way quantum information is organised in lower-dimensional space. Consider this again: a whole spatial dimension, something you can physically move through, measure, and even lose things in, is not part of the fundamental description. Instead, entanglement generates it. When I first understood this, I sat with it for a while. A whole spatial dimension, something you can move through, measure, build in, generated entirely by quantum information. I needed a moment. If one dimension can emerge from quantum information, why not the other two, or all of spacetime for that matter? The question that AdS/CFT forces us to ask is this, and the answer is beginning to reshape physics.

To understand AdS/CFT, we need to discuss Anti-de Sitter space, which is often shortened to AdS. General relativity permits various types of spacetime, depending on the cosmological constant. Spacetime can be flat, which means it has zero curvature. De Sitter space, describing an accelerating universe, has positive curvature. Anti-de Sitter space, on the other hand, has negative curvature, curving like a saddle in all directions. AdS is a solution to Einstein's equations that has a negative cosmological constant. It can be thought of mathematically as a hyperboloid in a higher-dimensional flat space, but you do not have to think about this in detail. The properties are what matter: negative curvature is everywhere. Space curves in a way that is similar to a saddle, showing negative curvature in every direction. This is different from our universe, which is nearly flat or slightly positively curved. A boundary at infinity. Unlike flat spacetime, where infinity is at an infinite distance, AdS has a timelike boundary that is finite.

Light can reach and bounce back from this boundary in finite time, which is where the physics takes place. AdS has a large symmetry group, which is the isometries of hyperbolic space. These symmetries correspond precisely to the symmetries of the dual quantum theory. Focus on five-dimensional AdS—AdS_5. Including time, this spacetime has four spatial dimensions and one time dimension. The boundary of this space is four-dimensional, with three spatial dimensions and one time dimension. Maldacena's insight is that the bulk, which is the five-dimensional interior of AdS, is not necessary. Everything happening in this five-dimensional space, including gravity, can be fully described by a four-dimensional quantum theory that exists on the boundary. The five-dimensional spacetime emerges from four-dimensional quantum physics.

What exists on the boundary? A special type of quantum field theory, known as a *conformal field theory* (*CFT*). "Conformal" means the theory looks the same at all length scales. Whether you zoom in or zoom out, the physics remains the same. There is no intrinsic ruler or fundamental length. When you move along the boundary's energy scale zooming in (higher energy, ultraviolet) or zooming out (lower energy, infrared) you are, in the gravitational language, moving inward or outward through the bulk. This scale invariance is a powerful symmetry that tightly constrains the theory. In Maldacena's original construction, the specific CFT is the N=4 super Yang-Mills theory in four dimensions. The theory is related to the one describing the strong nuclear force, QCD, but has more supersymmetry and a different matter content. Super Yang-Mills is a quantum field theory of gauge bosons, like photons but non-abelian, and fermions. The interactions are governed by gauge symmetry, which is SU(N) for large N. It is purely quantum mechanical, with no spacetime curvature, no metric field, and no gravity. This theory exists on the boundary of AdS, the four-dimensional flat space, which is three dimensions of space plus time. It is just quantum fields evolving according to the rules of quantum field theory. The boundary quantum field theory is exactly the same as gravitational physics in the five-dimensional AdS bulk.

The AdS/CFT correspondence says exactly this: String theory on AdS_5

× S^5—five-dimensional Anti-de Sitter space times a five-dimensional sphere—equals N=4 super Yang-Mills theory on the four-dimensional boundary, which is flat Minkowski space. Every calculable quantity in the gravitational theory has a counterpart in the quantum field theory. States observables processes everything matches exactly. Not approximately, not in some limit. Exactly. Every quantum state in gravity has a matching state in the CFT. Every measurable quantity has its counterpart. The partition functions agree. The correlation functions agree. Every quantum state in one theory has a corresponding state in the other. And observables in gravity is to operators in CFT. Each measurable quantity has its counterpart. Here, partition functions match. The partition function in statistical mechanics, which encodes all thermodynamic information, matches on both sides. And correlation functions match. The correlations of fields in gravity are equal to the correlations of operators in CFT. This is an exact duality, a true equivalence. One theory has five spatial dimensions, featuring gravity, curved spacetime, black holes, and gravitational waves. The other theory has three spatial dimensions and one time dimension, with no gravity, flat spacetime, and quantum fields that have gauge symmetry. There are two descriptions of the same physics. To use the duality, you need a dictionary, a translation guide that helps you move between theories. The correspondence comes with a dictionary, and it is worth reading slowly because each entry is remarkable in its own right. Moving inward through the bulk deeper into the AdS interior; is equivalent to moving to higher energies in the boundary theory. Moving outward, toward the boundary, is equivalent to lower energies. The bulk's extra dimension is the boundary's energy scale, made literal. A field propagating through the bulk gravitational theory has an exact partner in the boundary: a specific quantum operator. Turn on a source for a boundary operator and you change the boundary conditions for a bulk field. These are not analogies. They are exact equalities, the same physical event described in two languages. A black hole sitting in the bulk corresponds to a boundary theory in a thermal state, with the black hole's Hawking temperature equal to the temperature of the quantum field theory. The entropy of the black hole equals the thermodynamic entropy of the thermal state. These matching relationships carry an almost uncomfortable precision: one theory in five dimensions, featuring gravity, curved spacetime, and black holes; the other in four dimensions, with no gravity, flat spacetime, and quantum fields governed by gauge symmetry. They are the same physics. Not approximately. Not in some limit. They compute identical answers for every question you can ask of both. Switching on a source for a CFT operator effectively changes the boundary conditions for a bulk field. Black holes in bulk is to Thermal states in CFT. A black hole in AdS corresponds to a finite-temperature state in CFT. The Hawking temperature of the black hole is equal to the temperature of the CFT thermal state. The entropy of the black hole is equal to the entropy of the CFT. Entanglement entropy in CFT is to minimal surfaces in bulk. This concept is known as the Ryu-Takayanagi formula, which was discovered by Shinsei Ryu and Tadashi Takayanagi in 2006. To understand this, take a region on the boundary and calculate its entanglement entropy in CFT. The result is equal to the area of the minimal surface in the bulk whose boundary matches that region. This relationship shows that entanglement in quantum theory constructs geometry in gravitational theory. This last correspondence is the linchpin as it's the bridge connecting holography to emergent geometry. The Ryu-Takayanagi formula is, in its quiet way, one of the most extraordinary equations in modern physics. It equates a quantity computed entirely from quantum correlations with a quantity defined by the geometry of spacetime. Entanglement, which Einstein dismissed as spooky action at a distance, turns out to be the very architect of the geometry he held most dear. Quantum entanglement on the boundary literally builds space in the bulk.

So why is it called holographic? The five-dimensional gravity in the bulk encodes the duality in the four-dimensional quantum field theory on the boundary, making it holographic. The radial direction in AdS isn't a fundamental aspect of the CFT, but instead emerges from the energy scale structure of the quantum theory, making one spatial dimension emergent. Think of it like this: the CFT is fundamental, a quantum field theory that exists in four-dimensional space. There's no fifth dimension. If you *consider* the theory's behaviour at different energy scales such as UV versus IR, or short distances versus long distances, this energy scale structure is translated into a geometric radial direction in the dual gravitational description. Specifically, the entanglement structure of the CFT determines the bulk geometry. Highly entangled regions of the CFT correspond to nearby regions in the bulk, while weakly entangled regions correspond to distant regions. Distance is equivalent to entanglement, and geometry is a quantum correlation. This is holography made precise.

Why We Trust It

You might be sceptical. How do we know that this duality is correct? We cannot prove it with complete certainty, as solving both theories exactly is impossible. But we have overwhelming evidence. The evidence is, I think, as close to overwhelming as physics outside mathematics ever reaches. AdS_5 and N=4 super Yang-Mills theory share identical symmetries, with the same conformal symmetry group, SO(2,4), and identical supersymmetry. The particle spectra of both theories, calculated at weak coupling, match exactly, with each state corresponding precisely to its counterpart. Correlation functions computed at strong coupling using gravitational descriptions align perfectly with weak-coupling results from perturbation theory, wherever direct comparisons are possible. Thermodynamic calculations also give the same results, and the black hole entropy from the Bekenstein-Hawking formula is identical to the entropy of the dual thermal state calculated using statistical mechanics, with no difference at all. Phase diagrams of black holes in AdS, including stability, evaporation, and phase transitions, accurately replicate the thermodynamics of the CFT across various temperatures and chemical potentials. The Ryu-Takayanagi formula, which establishes that entanglement entropy is equivalent to minimal surface area, has been rigorously validated through various examples and consistently delivers accurate results. The thermalisation timescale following a sudden perturbation in the CFT precisely matches the timescale of black hole formation in the bulk. Recent research clearly shows that the duality works like a quantum error-correcting code, where the CFT stores bulk information in a way that protects it from errors. Over 25 years and thousands of calculations have consistently supported this mathematical concept, and as a result, the duality rests on a strong foundation in theoretical physics, similar to other well-established results. AdS/CFT is more than a mathematical curiosity. It offers valuable lessons about spacetime and helps us gain fresh insights into the nature of the universe.

- Spacetime is emergent. In AdS, the bulk spacetime and its geometry, metric, and curvature, emerge from quantum entanglement in the boundary theory. CFT has no fundamental geometry; instead, geometry is derived, not fundamental.
- Gravity is emergent. CFT has no gravity; instead, it has quantum fields with gauge interactions. The dual description, however, includes gravity, with features such as Einstein's equations, black holes, and gravitational waves, which emerge from non-gravitational quantum physics.
- Dimensionality is emergent. CFT comprises four spacetime dimensions, consisting of three spatial dimensions and one time dimension. The bulk, on the other hand, has four large spatial dimensions and one time dimension, which constitute the AdS_5 part, as well as five additional compact dimensions that are curled up at the Planck scale, forming the S^5 part of $AdS_5 \times S^5$ in the complete string theory description. We will focus on the five large dimensions of AdS_5 for our analysis. This extra dimension emerges from the RG flow, which is the energy scale structure of quantum theory.
- Black hole interiors are holographic. The black hole interior, which is behind the horizon, is encoded in the thermal correlations at the boundary. It is reconstructed from the information at the boundary. The interior is not fundamental. Again, all these lessons apply to AdS/CFT, but they also suggest a universal principle. So, does this

apply to our universe? The catch is that our universe is not AdS. We live in a nearly flat spacetime that has a small positive cosmological constant. Our universe is accelerating, rather than decelerating, which is what you would expect in AdS. Unlike AdS, our universe does not have a special boundary. Can AdS/CFT tell us anything about reality? Directly, no. Maldacena's duality is specific to Anti-de Sitter geometry, which has a special boundary. Our universe does not have this structure. But the principle might be generalised. Physicists are actively exploring:

- De Sitter holography. Our accelerating universe is similar to de Sitter space. Does it have a holographic dual? The cosmological horizon could act as a holographic screen, but building the dual theory is more complicated than in AdS. Some progress has been made, but a complete duality has not yet been found.
- Flat space holography. We can try to formulate holography in ordinary flat (Minkowski) spacetime. Some proposals are to describe bulk physics via scattering amplitudes at null infinity or by encoding data on the celestial sphere. It's speculative, but the research is promising.
- Cosmological holography. Can we describe the Big Bang, inflation, and cosmic microwave background holographically? Some hints suggest the cosmic microwave background might have a holographic structure, but this remains unconfirmed.

The verdict is out. However, the success of AdS/CFT in its domain suggests that holography is a fundamental principle, rather than just a quirk of AdS geometry. The geometry of Anti-de Sitter space curves inward. It is like a universe with a negative cosmological constant. On the other hand, our universe has a small positive cosmological constant, expanding outward, with a geometry that resembles a sphere rather than a saddle. These two geometries represent substantial differences, as they feature radically distinct boundary structures. AdS has a fixed timelike boundary where the CFT exists, while de Sitter does not have this feature. The whole mathematical framework of AdS/CFT, including the dictionary, the RT formula, and the precise equivalence, relies on that boundary. Translating it to de Sitter space is not simply about tweaking parameters but involves rebuilding the whole framework, which remains incomplete. Scientists know this challenge as the de Sitter problem, and they consider it one of the most intriguing open questions in theoretical physics. This is important for our exploration because we argue that holography provides evidence that geometry emerges from entanglement. Researchers find the strongest evidence in AdS, while they are extending this idea to our own universe, de Sitter, which is still a work in progress. The principle appears strong enough to adapt, and there are ongoing efforts in de Sitter holography.

Distance Is Entanglement

Now, let's take a closer look at the Ryu-Takayanagi formula, as it is to understanding how entanglement builds geometry. *Consider* the CFT on the four-dimensional boundary. We divide it into two regions: one is region A, and the other is everything else. The total quantum state of the full CFT is pure, it has zero entropy. However, when we look only at region A and trace out everything else, the reduced density matrix becomes mixed, and it has entropy. This is what we call entanglement entropy:

$$S_A = -\mathrm{Tr}(\rho_A \log \rho_A)$$

It measures how entangled region A is with the rest of the system. A high entanglement entropy means there are strong correlations between A and its complement. Now, look at the AdS bulk. The minimal surface in the bulk spacetime, whose boundary is ∂A, the boundary of region A on the CFT boundary, needs to be found. This surface is called the Ryu-Takayanagi surface. To find the area of this surface, calculate it in Planck units. The RT formula is as follows:

$$S_A = \mathrm{Area}(\text{RT surface}) / (4G_N)$$

G_N here is Newton's gravitational constant.

Entanglement entropy in boundary quantum theory is equal to the geometric area in bulk gravitational theory. This is impressive. The quantum information (or entanglement) on one side is equal to the geometric area on the other. Entanglement shapes geometry. Study a concrete example. Suppose the boundary is a circle and region A is a segment of that circle. The calculation of entanglement entropy depends on the size of the segment and the details of the CFT state. In the bulk, the minimal surface anchored at the segment's endpoints forms a curve that dips into the bulk. Calculate its length, which is the area in one dimension. The RT formula states that these two quantities match entanglement entropy is equal to curve length. They are equal. If you change the CFT state by heating it to a finite temperature. The entanglement entropy too changes. In the bulk, a finite temperature means that a black hole appears, which changes the bulk geometry and the minimal surface. However, the RT formula still holds. This has been verified in countless examples, including different regions, states, and geometries and it consistently works.

The RT formula shows something profound: the entanglement at the boundary decides the distance in the bulk. Entanglement and geometry are directly proportional: highly entangled boundary regions correspond to short minimal surfaces (close bulk connections), while weakly entangled regions correspond to long ones. This relationship is precise: the distance in the bulk directly relates to the entanglement structure on the boundary. You can reverse this logic. Begin with the boundary theory, a quantum system that has some entanglement pattern. The bulk geometry can be reconstructed from the entanglement structure alone, which is made up of pure quantum information. The web of quantum correlations determines the spatial distances. This has been formalised in *tensor network* models. The CFT state is represented as a tensor network, a mathematical structure that encodes entanglement. The geometry of the network, which is how the nodes connect, determines the emergent bulk geometry. In simple cases, spacetime can be seen to emerge from entanglement. You start with qubits, organise their connections a certain way, and then geometry appears. This captures the concept of emergent spacetime: geometry is not fundamental, but rather it arises from the structure of quantum information. AdS/CFT is the most obvious example of holography and emergent spacetime, but it is not the only one. Matrix theory, which includes M-theory, reduces to a model of interacting matrices in certain situations, and spacetime emerges from these matrices. Celestial holography is similar, but it works differently, encoding physics on the celestial sphere and expressing scattering amplitudes as correlators. Even without using string theory, tensor network models show that spacetime can emerge from simple things like qubits on a lattice, where entanglement creates spacetime. Recently, holographic quantum error correction has found a strong link between holography and error-correcting codes, where the bulk is the logical space and the boundary is the physical space. Here lies the clue that, information in the bulk is safely encoded across the boundary, protecting it from errors, and explains why bulk reconstruction is possible and how entanglement helps build geometry.

All these approaches lead to the same conclusion: spacetime emerges from quantum information. Now, let's look at dimensionality itself. In AdS/CFT, the bulk has an extra spatial dimension compared to the boundary. The boundary CFT exists in a four-dimensional spacetime, which has three dimensions of space and one of time. The bulk AdS exists in a five-dimensional spacetime, which has four dimensions of space and one of time. The extra dimension is not fundamental; it emerges from the structure of energy scales, or the RG flow, in the quantum theory. If one dimension can emerge from quantum information, then why not more? Why not all of them? Suppose that spatial dimensions do not exist at a fundamental level, that we only have abstract quantum degrees of freedom, and that spacetime and all its dimensions arise from the entanglement of these degrees of freedom. This is the framework. AdS/CFT has proved that one dimension can emerge from quantum information. The challenge now is to show that all three spatial dimensions can emerge in a similar way. Part III makes the case that three-dimensional space emerges from the quantum entanglement structure, as three is the natural dimensionality for realistic entanglement networks under physical constraints. But we need one more piece. We need to understand what entanglement is, how to measure it, and its relation to geometry.

Let's review where we are. Einstein's general relativity describes gravity as the curvature of spacetime, presenting an elegant and successful concept, but it assumes that spacetime is fundamental. Black hole thermodynamics has revealed cracks in this theory. The area law we have traced since Chapter 5 reappears here, suggesting that two-dimensional information encodes three-dimensional space. AdS/CFT made this precise: five-dimensional gravity in AdS is equal to four-dimensional quantum field theory on the boundary. One dimension emerges from the quantum information. The Ryu-Takayanagi formula has a role: it states that entanglement entropy is equal to the minimal surface area. The entanglement at the boundary builds up distance in the emergent bulk geometry. Now we are looking at a more general situation. If one dimension emerges in AdS/CFT, can the same thing happen with all three spatial dimensions in a different setting? Can we begin with just quantum information, without any geometry or dimensions, and end up with a three-dimensional space? The answer is yes. Naturally, three dimensions emerge as the structure of entanglement networks when realistic constraints are applied. This is Part III. We will build space from qubits and explore why specific entanglement patterns result in three-dimensional geometry, rather than two or four dimensions. We will understand why our spatial reality is three-dimensional. But first, we need to understand quantum entanglement itself, not just as an abstract connection between quantum systems, but as the concrete foundation on which geometry is built.

AdS/CFT does not literally describe our universe, we do not live in negatively curved space. But it demonstrates what is possible: one spatial dimension emerging from quantum information in a rigorous, extensively tested theory. That proof of principle is our roadmap. If one dimension can emerge, why not all three? That is Part III. We will take entanglement seriously as the substance from which space is composed, explain why three-dimensional geometry emerges naturally from realistic entanglement networks, connect these ideas to experiments, and consider what it means if reality is fundamentally informational. By the end of Part III, you will see why space has three dimensions.

III

Architecture of Space

"There are more things in heaven and earth, Horatio, than are dreamt of in your philosophy." — Hamlet

More entanglement, shorter distance.
Less entanglement, further apart.
Geometry is not given. It is earned.

7

Quantum Connection

Why does measuring one particle have an instant effect on another, even when they are light-years apart? For decades, this question seemed to have only two possible answers, both of which were disturbing. Either quantum mechanics was wrong, or it was incomplete, missing some secret variables that would restore sensible local physics. Or nature really does permit instantaneous influences across cosmic distances, somehow skirting the light-speed limit set by relativity. Einstein preferred the *first* option. In 1935, he, along with his colleagues Boris Podolsky and Nathan Rosen, proposed a famous thought experiment. This involved two particles that interact and then separate to opposite ends of the galaxy. If you measure one, you instantly know the state of the other. How is this possible? Einstein argued the only sensible explanation was that both particles had predetermined properties all along—hidden variables that quantum mechanics had failed to describe. Thirty years later, physicist John Bell shattered this hope. He mathematically proved that no theory with local hidden variables could match the predictions made by quantum mechanics. Nature was faced with a choice: it had to either give up locality, allowing for instantaneous influences, or give up quantum mechanics altogether. The experiments we have done have chosen for us. Beginning in the 1970s and continuing through to increasingly sophisticated tests today, every measurement has confirmed quantum mechanics. Local hidden variables are ruled out. The instantaneous correlations are real. This phenomenon is called quantum entanglement, and it is not a flaw in our understanding of reality. It is the foundation of reality itself. That last sentence is not just for effect; it really sums up the main point of this book. Everything that follows—geometry, dimensions, and the number three—stems from taking entanglement seriously as the foundation of space.

What Is Entanglement? If you strip away the philosophy and history, entanglement boils down to a simple mathematical fact: some quantum states just cannot be factorised. Start with a single electron. Its quantum state is described by a wavefunction ψ , a mathematical object that encodes all measurable information about the particle. Before measurement, the electron exists in a superposition, meaning it is simultaneously spin-up and spin-down, with each possibility having a probability amplitude. When you measure the spin, you get a definite answer, either up or down. The superposition then collapses, according to the Copenhagen interpretation, or the universe branches, as per the many-worlds interpretation. In either case, after measurement, the electron occupies a definite state. Now *consider* two electrons. If they are independent, their combined wavefunction is simple: $\psi\_total = \psi_1 \otimes \psi_2$. Each particle has its own state, and together they form a product state. The particles maintain their separate identities. But quantum mechanics permits something even stranger: the total wave function can become entangled, and this cannot be expressed as a product of individual states. We have two spin-1/2 particles, and we are looking at their singlet state:

$$|\psi\rangle = (1/\sqrt{2})(|\uparrow\rangle_1|\downarrow\rangle_2 - |\downarrow\rangle_1|\uparrow\rangle_2)$$

Read this carefully. The state is a superposition, meaning the first particle might be spin-up and the second spin-down, or the first spin-down and the second spin-up. The particles are anti-correlated. But this is important since each particle does not have a definite spin on its own. Although the spin values of the two particles are perfectly correlated, they each exist in an indefinite state. This cannot be factorised. It is not possible to express it as the state of particle 1 multiplied by the state of particle 2. The properties of the particles are linked together. This is what is meant by entanglement. Now separate the particles. Put one in a lab in the Indian Institutes of Science Education and Research, Pune and the other on a spacecraft bound for Alpha Centauri, the closest star system to our Solar System. The entanglement persists because distance is irrelevant. Measure particle 1's spin along the vertical axis. You will get a random result, say spin-up. The moment you make this measurement, particle 2's state becomes definite, it has spin-down along the same axis. When you measure particle 2, you will confirm the anti-correlation. This holds true even across light-years. The correlation is instantaneous, perfect, and has been experimentally verified thousands of times. However, no signal travels between the particles. Entanglement does not allow for faster-than-light communication, which we will see shortly. The correlation between the particles is real, but it is not a cause-and-effect relationship in the usual sense. Instead, it is more complex than that. This is what we mean by entanglement: quantum correlation that cannot be reproduced by any classical mechanism. Einstein, Podolsky, and Rosen considered entanglement to be proof that quantum mechanics was incomplete, and their argument was deceptively simple. Suppose you measure particle 1 and find it has spin-up. Immediately, particle 2 must have spin-down, even if it is light-years away. But how does particle 2 "know" what its spin is? Two possibilities:

- A. Particle 1 sends a faster-than-light signal to particle 2 when measured, forcing it into the spin-down state. This violates relativity.
- B. Particle 2 already possessed spin-down before it was measured. The measurement only revealed a pre-existing property, a "hidden variable" that quantum mechanics cannot describe.

EPR chose option B. They concluded that quantum mechanics is incomplete. According to them, particles carry definite properties at all times; the theory just doesn't account for them. The wave function only gives statistical information about the underlying reality. It was a reasonable argument, but it was also wrong.

In 1964, physicist John Bell put forward a mathematical proof that challenged the EPR argument, and his theorem has remained one of the most profound findings in physics. Bell attended Queen's University Belfast as an undergraduate and later worked at CERN. He showed that any theory having local hidden variables, where particles have predetermined properties and no influences occur faster than light, must follow certain mathematical constraints on correlations between measurements. These constraints are known as Bell inequalities. He then proved that quantum mechanics does indeed violate these inequalities. The conclusion is stark: to measure entangled particles, do so along various axes. If local hidden variables exist, the correlations must satisfy Bell's inequality. If quantum mechanics is correct, the correlations will violate it. Experiments began in the 1970s, with increasingly sophisticated tests being carried out throughout the 1980s. Alain Aspect's experiments near Paris were particularly decisive. Every test violated Bell's inequality, confirming quantum mechanics and ruling out local hidden variables. What does this mean? Nature is non-local. Entangled particles show correlations that cannot be explained by any local theory. Every popular physics book explores the strangeness of entanglement, including its spookiness, thought experiments, and Bell violations, but this is just the introduction. What these books often fail to explain is how entanglement behaves when you have a vast number of particles, approximately 10^{23}, all entangled in complex patterns governed by the laws of physics. This is where our story truly starts, focusing not on the oddity of two entangled particles, but on the intricate architecture of many particles. The entanglement of two particles is a fascinating phenomenon, but it is the entanglement structure of a many-body quantum system that forms the foundation of reality.

Before your imagination runs wild with schemes for instantaneous communication, let us understand why entanglement cannot transmit information faster than light. When you measure particle 1, you get a random outcome: a 50% probability of spin-up and a 50% probability of spin-down. You have no control over the result. At the same time, particle 2 becomes correlated with your measurement, but its outcomes are still equally random. The correlation emerges only when you compare results. When you measure particle 1 along the vertical axis and then particle 2 along the same axis, you'll find they have a perfect anti-correlation. But if you look at each individual measurement sequence, it looks completely random. *Imagine* Alice measures particle 1 and obtains: up, down, up, up, down. Bob measures particle 2 and obtains: down, up, down, down, up. The sequences are perfectly anti-correlated. However, neither Alice nor Bob notice anything unusual in their individual data. The pattern emerges only when they compare notes. This comparison needs classical communication, where Alice sends her results to Bob, or Bob to Alice. This communication is limited by the speed of light. So while entanglement creates non-local correlations, it does not allow for faster-than-light signalling. The correlations remain hidden until classical information brings them together. This is why entanglement does not violate relativity. No information can travel faster than light. The non-locality exists in the correlations themselves, not in any causal influence.

Measuring It

How do we measure the extent to which two systems are entangled? For pure states of subsystems A and B, we use entanglement entropy. We begin with the total system AB in a pure state, $|\psi\rangle$. Because the combined system is pure, it has zero entropy, meaning it is in a definite quantum state with no uncertainty. Now, *consider* subsystem A alone. We describe A independently by "tracing out" B's degrees of freedom, which gives us the reduced density matrix:

$$\rho\_A = Tr\_B(|\psi\rangle\langle\psi|)$$

If A and B are unentangled, ρ_A remains pure, meaning $\rho\_A = |\psi\_A\rangle\langle\psi\_A|$. In this case, the von Neumann entropy $S\_A = -Tr(\rho\_A \log \rho\_A)$ equals zero. However, when A and B are entangled, ρ_A becomes mixed - it is a statistical combination of different pure states. As a result, ρ_A has non-zero entropy. This is quantified by the von Neumann entropy:

$$S\_A = -Tr(\rho\_A \log \rho\_A)$$

The maximum entanglement occurs when ρ_A is maximally mixed, with all its eigenvalues being equal. At this point, S_A = log(dim A), with dim A being the dimension of A's Hilbert space. For two qubits in the singlet state, the entanglement entropy is $S = \log_2 2 = 1$ ebit (one bit of entanglement), which is the maximum for two qubits. Throughout this book, entanglement entropy is measured in bits using log base 2. The entanglement entropy has a notable property: for pure states of the total system, S_A = S_B. Whether you trace out A or B, the entropy is the same. This reflects a fundamental truth: entanglement is a property of the relationship between A and B, rather than of either subsystem on its own. Let me clarify what entanglement is and what it is not. Entanglement is a quantum correlation, a correlation that cannot be reproduced classically. Classical correlations do not violate Bell inequalities, but quantum entanglement does. You can have classical correlations without entanglement. Say you have two boxes: one box contains a red ball and the other box contains a blue ball. Ship these boxes to opposite sides of the planet. If you open one box, you instantly know the colour of the other box. This is classical correlation, and it is not entanglement. The difference is due to superposition.

The entangled state $(|\uparrow\rangle_1|\downarrow\rangle_2 - |\downarrow\rangle_1|\uparrow\rangle_2)/\sqrt{2}$ is different from a 50-50 classical mixture. This state is a coherent superposition, and it exhibits interference effects that violate the Bell inequalities. Classical mixtures cannot exhibit this behaviour. Mathematically, entanglement refers to a state that cannot be factorised. Operationally, this results in correlations that go beyond classical limits. Physically, it means that the systems share quantum information that neither one has on its own. In quantum information theory, entanglement is a resource, something that needs to be created, distributed, and then consumed. Quantum teleportation illustrates this. Alice and Bob have an entangled pair. Alice performs a joint measurement on her entangled particle and an unknown quantum state she wants to send to Bob. She transmits the classical measurement outcome to Bob. Bob uses this information to manipulate his entangled particle, thus transforming it into the original state Alice wanted to send. The quantum state effectively "teleports" from Alice to Bob. This is not science fiction teleportation, with no faster-than-light travel, and the original state is destroyed. But this is genuine quantum state transfer, with entanglement acting as the carrier. Quantum cryptography uses entanglement for provably secure communication. Entangled particles generate shared random keys. If anyone intercepts the communication, the entanglement is disturbed, revealing the intrusion. Quantum computing uses entanglement to achieve computational speedup. Algorithms like Shor's, which are used for factoring, and Grover's, which are used for search, exploit entanglement to outperform classical computation. Entanglement is the core ingredient, and without it, quantum information processing becomes classical computation. There is an important point that deserves more attention. Throughout this chapter, we have discussed entanglement as if its meaning is well understood. But quantum mechanics still faces a long-standing challenge known as the measurement problem. When we observe a quantum system, it seems to cause its superposition to collapse into a clear outcome. Yet the Schrödinger equation, which governs all quantum evolution, is completely reversible and does not show this collapse. This inconsistency needs to be addressed. The Copenhagen interpretation suggests that collapse is a real event, though it does not clarify when or why it happens. So the many-worlds interpretation argues that collapse never actually occurs, proposing instead that the universe branches into multiple paths, each reflecting a possible outcome. There are also other interpretations, like consistent histories, QBism, and relational quantum mechanics, which offer unique explanations. And currently, we lack experimental evidence to distinguish between them. This lack of agreement among physicists is not just a theoretical detail but something that influences our current understanding of quantum mechanics. Importantly, one of the main ideas that entanglement acts as a physical resource underlying geometry holds strong across all major interpretations of quantum mechanics. The correlations between entangled particles can be tested empirically, are mathematically sound, and do not rely on how one views wavefunction collapse. In the end, the differences between interpretations relate more to their deeper philosophical meanings rather than the physical principles upon which this book is based. But entanglement is fragile. Environmental interactions, or *decoherence,* destroy it. This is why building quantum computers is hard: it is an engineering nightmare to maintain entanglement while protecting it from environmental noise. Still, under the right conditions like low temperatures and well-isolated systems, entanglement can survive. This has led to the development of quantum engineering, which involves the deliberate creation and manipulation of entangled states for technological applications. Entanglement has a peculiar property that is monogamy. Say if particle A is maximally entangled with particle B, it cannot be entangled with particle C. This means that entanglement is exclusive. Generally, the greater the entanglement between A and B, the weaker the entanglement with other things will be. This is because quantum correlations always involve trade-offs. The monogamy inequality expresses this mathematically. For three parties A, B, and C, Coffman, Kundu, and Wootters proved in 2000 that:

$$E(A{:}BC) \geq E(A{:}B) + E(A{:}C)$$

Here, E(A:B) denotes an entanglement measure (specifically, the squared concurrence) between A and B. Unlike classical correlations or mutual information, which can be freely shared among many parties, entanglement measures obey this strict inequality; a qubit's entanglement is a finite budget that must be divided among its partners. But quantum entanglement obeys different rules. It limits how entanglement can be shared in quantum networks. It influences the structure of quantum states in many-body systems. And, as we will see, it greatly restricts the geometries that can result from entanglement. See it as a resource allocation challenge. You have a limited budget, which is your total entanglement capacity, and you must allocate it across several connections. If you distribute it too thinly, no single connection will be strong enough, but if you concentrate it, you will lose connectivity. The optimal allocation hinges on the network topology, which is where dimensionality plays a central role, three-dimensional networks allocate entanglement differently than two-dimensional or four-dimensional ones, despite having the same budget constraints, as the geometry fundamentally changes everything. In emergent spacetime, monogamy limits the possible entanglement patterns. Every configuration of quantum correlations cannot exist. The allowed patterns, which are consistent with monogamy and other quantum information constraints, determine the geometries that can emerge.

Entanglement in Many-Body Systems

Most quantum systems that we study contain a large number of particles, typically 10^{23} particles in a macroscopic solid, or infinitely many degrees of freedom in a quantum field. In these many-body systems, the entanglement structure becomes rich and complex. Different regions exhibit varying levels of entanglement. You might have short-range entanglement, where only nearby degrees of freedom are entangled, or long-range entanglement, where distant parts of the system remain correlated. In any dimension, ground states of local Hamiltonians display area-law entanglement, where entropy is tied to the boundary area of a region, not its volume. The area-law structure has a decisive role in the Ryu-Takayanagi formula and emergent geometry. While the area law holds in 2D, 3D, and 4D systems, three dimensions have distinct properties when area-law states are subjected to stability analysis, error correction, and entanglement monogamy constraints, which we explore in Chapters 10 and 11. The entanglement entropy scales with the boundary area, not the volume. Excited states or systems at quantum phase transitions can exhibit volume-law entanglement, where entropy scales with the volume of the region, indicating the presence of long-range correlations. Thermal states at finite temperature have extensive entropy, with both entanglement and thermal entropy scaling with volume. The separation of thermal and quantum contributions is still an active area of research. What matters is that the entanglement structure in many-body systems isn't random. It's shaped by dynamics, locality, and conservation laws, which create specific architectures of quantum correlation. These patterns will help us understand emergent geometry.

How do you mathematically represent the entanglement structure of a quantum state? For small systems, you can write the full wave function. For a system with two qubits:

$|\psi\rangle = \alpha|00\rangle + \beta|01\rangle + \gamma|10\rangle + \delta|11\rangle$, this has four complex coefficients. But this approach does not scale well. An N-qubit system needs 2^N complex numbers. For example, if N = 300, the number of complex numbers required exceeds the number of atoms in the observable universe. Fortunately, most physically relevant states possess structure, which includes locality, limited entanglement, and area-law scaling, and this structure can be efficiently captured by tensor networks. In a tensor network, nodes represent tensors, which are multidimensional arrays of numbers. Edges between nodes represent indices that need to be contracted, or summed over, to give the full quantum state. The topology of the network, how the nodes connect to each other, shows the entanglement structure.

Matrix Product States (MPS) provide an efficient representation of one-dimensional quantum systems. Each site has a tensor with three indices: one physical index for the local quantum state and two virtual indices that connect to its neighbours. When the virtual indices are contracted, they yield the wave function. MPS can efficiently capture ground states of one-dimensional local Hamiltonians, as those with area-law entanglement compress naturally into MPS form. MERA, the Multi-scale Entanglement Renormalisation Ansatz is where tensor networks started looking less like a computational convenience and more like a clue about nature. Its structure is hierarchical: you start with a boundary of qubits, apply a layer of entanglement operations that coarse-grain the system, then apply another, then another, each layer encoding correlations at a different scale. The result is a tree of entanglement with depth. And that depth that extra direction that MERA naturally grows into looks exactly like the radial dimension of Anti-de Sitter space. The boundary of AdS is where the CFT lives. The interior of AdS is where gravity lives. MERA says: if you want to efficiently represent the quantum state of a conformal field theory, you build a structure whose geometry is AdS. You do not choose AdS. It emerges from the requirement of efficient entanglement. Vlatko Vedral, when I read his account of this, put it in a way I have not been able to improve on: the universe may be a vast computation, and space may be its memory structure. The way the nodes in a network graph are connected, which is how they link to each other, defines a spatial structure that emerges from this. If you change the network topology, you also change the geometry because the pattern of quantum correlations shapes space.

I have devoted this chapter to entanglement as it is the key to understanding dimensionality. In Part I, I explained why three dimensions are special as they allow for stable orbits, robust atoms, and topological advantages. But, I didn't explain why our universe has three dimensions. The answer can be in the entanglement structure. Quantum systems, where degrees of freedom mainly interact with their neighbours, exhibit characteristic entanglement patterns. These patterns have a distinct mathematical structure, characterised by area-law scaling, a finite correlation length, and a hierarchical organisation. These structural constraints decide which geometries are possible. Think of it this way. In one dimension, every qubit has exactly two neighbours. The entanglement network is a chain. You can build correlations along the chain, but you cannot build a hierarchy; there are no shortcuts, no loops, no way for distant parts of the system to communicate without going through every qubit in between. The geometry is too thin to support anything complex. Two dimensions gives you more room, but not enough: the boundary-to-area ratio is wrong, the entanglement budget runs out before you can encode anything interesting, and the topological constraints that knot theory demands cannot be satisfied. Then in four dimensions, the opposite problem. Too many connections. Perturbations do not damp out; they propagate and amplify. The entanglement structure is not stable, small changes in one part of the network cascade globally. Three dimensions is the case where both failure modes are avoided. It is the only integer where the network is richly connected without being catastrophically over-connected. Whether that is a coincidence or a theorem is the question the rest of this book is trying to answer. In Chapters 8-11, I'll develop the argument that entanglement builds geometry, that the structure of quantum correlations determines dimensionality, and that realistic quantum systems lead to three dimensions emerging. Particles do not just interact when they are close; they also remain connected through quantum correlations, even over huge distances in space. The existence of these connections is a reality, they can be measured, and classical physics cannot account for them. And these connections build space itself. Similarly, entanglement, or this quantum connection, is more than just a correlation between particles. It forms the basis from which geometry emerges. The patterns of quantum entanglement give rise to distance, dimensionality, and the structure of spacetime. In the next chapter, we will see this in clear terms. We will see how geometry emerges from entanglement. We will derive connections between quantum information theory and geometric structure, and understand how distance arises from correlation. Finally, we will look at the question of why we have three dimensions.

8

The Breakthrough

Suppose I give you a list of numbers, one for each possible way to divide a quantum field into regions. Each number indicates how entangled that region is with everything outside it. *Is it possible to reconstruct the geometry of spacetime using just these numbers?* The answer is surprisingly yes, and there is a precise recipe for doing so. Think of it as an assumption. The fact has been demonstrated clearly in controlled examples. If we start with quantum information and follow the formula, geometry emerges from pure correlation. We met the Ryu-Takayanagi formula in Chapter 6. Now let us use it as a blueprint for building space from scratch. Take your boundary region A. The boundary of A, ∂A, is a curve in d=2 or a surface in d=3 that sits at the edge of AdS space. Next, extend this boundary into the bulk and look for a surface in the bulk spacetime that: This surface must satisfy two conditions: its boundary must match ∂A, and among all surfaces meeting that requirement, it must have the smallest possible area, it is minimal. This is the Ryu-Takayanagi surface, denoted γ_A.

Imagine a soap film. Take a wire loop to serve as the boundary ∂A. Dip this loop into soapy water, and the resulting film will form a minimal surface. This surface has a shape that minimises surface tension, and it does so by having the smallest possible area. In bulk spacetime, the same principle applies: given a boundary ∂A, there is a unique surface with the smallest area. The area of this surface, as measured by the bulk metric, is a geometric quantity and it's an intrinsic property of the bulk spacetime geometry. Let's make this concrete. *Consider* a 2D CFT on the boundary, one spatial dimension plus time. When we freeze time at a moment and look at the spatial slice, the boundary is a line. Choose region A as an interval: $x \in [x_1, x_2]$. The boundary of ∂A is made up of two points: x_1 and x_2. In the bulk, you have 2D spatial geometry, with time frozen. The bulk is the upper half-plane: the region where $z > 0$ in the (x, z) plane. In this half-plane, the minimal surface γ_A connecting boundary points x_1 and x_2 forms a curve. What shape does it take? For empty AdS (the CFT vacuum), the minimal curve is a semicircle. It dips into the bulk, reaching its maximum depth at $z\_max = (x_2 - x_1)/2$, and then returns to the boundary. The length of this semicircle, which can be thought of as its "area" in one dimension, can be calculated using the AdS metric:

$$\text{Length}(\gamma\_A) = (c/3) \log[(x_2 - x_1)/\varepsilon]$$

where c is the central charge of the CFT (a constant characterising the theory), and ε is a short-distance cutoff. Now calculate the entanglement entropy of the same interval in the CFT vacuum:

$$S\_A = (c/3) \log[(x_2 - x_1)/\varepsilon]$$

The formulas match: S_A = Length(γ_A)/(4G_N), where G_N is Newton's constant in the bulk. Entanglement entropy equals minimal surface area divided by 4G_N. The entanglement entropy is equal to the minimal surface length, properly normalised. Both the quantum calculation and the geometric calculation yield the same outcome. Stop. Consider what we have just shown. We calculated the amount of entanglement between two regions, a quantum property, and the length of a curve through spacetime, a geometric property. What's eye-catching is that these two numbers are identical, not almost the same, but exactly the same. This suggests a deep connection between the two: entanglement and geometry are essentially the same thing. We began with a list of numbers, one for each way of dividing a quantum field into regions. We then followed a formula. And geometry appeared: not as an assumption, not as a background, but as a consequence. Distance is entanglement. Area is correlation. Space is what quantum information does when left to settle.

How It Was Derived

Ryu and Takayanagi derived this formula, but how did they do it? They used the replica trick from quantum field theory. The method calculates entanglement entropy by first computing Rényi entropy, a related but simpler quantity, then taking the limit as the parameter $n \to 1$. On the CFT side, we compute correlation functions on an n-sheeted Riemann surface, gluing the sheets together along region A. As n tends to 1, this yields the entanglement entropy. On the bulk side, the correlation function is the same as that obtained by evaluating the bulk path integral over a geometry that contains the minimal surface γ_A, appearing n times. As n approaches 1, the minimal surface area dominates the path integral, which gives:

$$S_A = \text{Area}(\gamma_A) / (4G_N)$$

The derivation is technical, involving complex analysis, saddle-point approximations, and careful limits. However, the logic is clear: CFT entanglement and bulk geometry are two ways to calculate the same thing, and they are related by AdS/CFT duality. The RT formula isn't something that's imposed; it's a natural consequence that follows from the duality. For more than two decades, the Ryu-Takayanagi formula has undergone extensive testing under a vast array of conditions, and it has emerged successful every time. The formula's validity extends to entanglement regions of every conceivable shape, whether they are intervals, strips, spheres, disconnected regions, or even regions with fractal boundaries. It has been shown to hold for all types of states, including vacuum, thermal, excited, and time-dependent configurations, with quantum corrections allowing it to precisely capture excited states. The formula holds across 2D, 3D, and 4D CFTs with no sensitivity to dimensionality of the boundary theory, and reinforcing its universal applicability. The formula inherently satisfies the fundamental mathematical constraints that define a well-established notion of entropy, including the inequalities of strong subadditivity, which emerge naturally from the geometric properties of minimal surfaces. This geometric framework automatically enforces quantum constraints without requiring external imposition. Furthermore, the formula integrates causality, as the information flow on the CFT side accurately aligns with the causal structure of the bulk, where spacelike-separated regions in the bulk correspond to commuting boundary operators. The RT formula has been refined systematically, going beyond its classical form with higher-order quantum corrections, all of which have been calculated independently and verified to match CFT predictions precisely. The formula also underlies subregion duality, which is the notable fact that a specific part of the boundary corresponds to a specific part of the bulk, a connection that is made possible because entanglement is a key aspect of the geometry, influencing its structure.

Every test has confirmed the formula. Despite twenty years of research and thousands of papers, no counterexample has been found. The RT formula is as firmly established as any result in quantum gravity. The original RT formula applies to static situations that are time slices of both bulk and boundary. What happens in dynamic cases? In 2007, Veronika Hubeny, Mukund Rangamani, and Tadashi Takayanagi generalised the formula for time-dependent scenarios, resulting in the HRT formula. For a boundary region A at a given time, the entanglement entropy is equal to the area of an extremal surface in the bulk, a surface that is minimal in space and extremal in spacetime, which allows it to evolve with time. This extremal surface is similar to a minimal surface, but in spacetime. It extremizes area, subject to constraints from the bulk geometry and the causal structure. The HRT formula reduces to RT in static cases, but it handles dynamics, including time-dependent geometries, black hole evaporation, and thermalisation processes. Further generalisations extend the framework in several important directions. The Quantum RT formula incorporates quantum corrections from bulk quantum fields,

giving S_A = (Area + quantum corrections)/(4G_N).

These corrections are essential for resolving the black hole information paradox. Building on this, researchers working in cosmological settings have introduced the concept of "quantum extremal surfaces" and "islands," referring to regions of the bulk that unexpectedly contribute to boundary entanglement. This recent development, captured in the island formula, has revolutionised our understanding of the information paradox in ways that continue to reverberate through the field. Finally, the entanglement of purification goes beyond entanglement entropy altogether, encompassing other information measures that each carry their own geometric interpretation in the bulk.

These generalisations strengthen the connection between quantum information and geometry. As we test more, the relationship proves to be deeper. The RT formula states that entanglement entropy is equal to area, but it also implies something more: the entire pattern of entanglement determines the complete metric of the bulk. To do this, *consider* multiple regions A_1, A_2, A_3, ... on the boundary. For each of these regions, the entanglement entropy S_A_i is calculated. Each of these entropies corresponds to a minimal surface area in the bulk. These minimal surfaces aren't independent. They're embedded in a single bulk geometry. The change in areas when the regions are varied constrains the bulk metric. It's like triangulation; knowing enough distances between points allows you to reconstruct the geometry. In the same way, knowing the minimal surface areas for different regions allows you to reconstruct the bulk metric. Here is the statement, plainly. If you know the entanglement entropy of every possible region on the boundary; every interval, every sphere, every disconnected combination, you can reconstruct the geometry of the bulk. Not approximately. You can derive the metric: the distances between points, the curvature of space, the shape of the gravitational field. The entanglement is not a probe of the geometry. The entanglement is the geometry. There is ambiguity deep in the bulk, far from the boundary, where the reconstruction becomes underdetermined. But near the boundary, the reconstruction is exact. Two different quantum states on the boundary, with different entanglement structures, correspond to two different bulk geometries, different physical spacetimes. This is what 'space from entanglement' means, stated in the most literal possible way: the pattern of quantum correlations in a lower-dimensional theory encodes the metric of a higher-dimensional space. If you want to understand why space has three dimensions, you need to ask what entanglement patterns can produce stable bulk geometries, and in how many dimensions those geometries can be self-consistent. That question is what Chapter 10 is for. Let's make the concept of distance more precise. Take two small regions A and B on the boundary. We need to define their "distance" in the bulk. One method to do this uses mutual information:

$$I(A:B) = S_A + S_B - S_{AB}$$

Here, S_AB represents the entropy of the region that combines $A \cup B$. In the CFT vacuum, I(A:B) can be calculated directly. The RT formula on the bulk side gives a statement about minimal surfaces. In small, well-separated regions, mutual information decreases exponentially with increasing bulk distance:

$$I(A:B) \sim e^{-d(A,B)/\ell}$$

Here, d(A,B) represents the bulk distance between A and B, with ℓ being a characteristic length scale. High mutual information means there is a strong correlation, which in turn means a small distance. Low mutual information means the correlation is weak, which means the distance is large.

A stunning insight is that the radial direction in AdS, which is the extra dimension beyond the boundary, emerges from the energy scale structure of the CFT. In quantum field theory, different energy scales are associated with different physics. The high energy scale, UV, involves short distances and high frequencies, while the low energy scale, IR, involves long distances and low frequencies. The renormalisation group, describes how the theory changes when you vary the energy scale. It flows from UV to IR, connecting the high-energy and low-energy descriptions. In AdS/CFT, the RG flow is seen in the radial direction. As you move deeper into the bulk, by increasing z, you flow to lower energies, which is the IR region. Conversely, moving towards the boundary, by decreasing z to 0, means you flow to higher energies, which is the UV region. The radial coordinate is not a fundamental spatial dimension in the CFT; there is no "z direction" in the boundary theory. It emerges from the structure of the energy scale.

This is clearly stated in the relation: $z \sim 1/E$. Here, E is the energy scale. At low energy, or the IR, z is large, meaning it is deep in the bulk. At high energy, or the UV, z is small, meaning it is near the boundary. The extra dimension, the entire radial direction of AdS, is emergent. It is not built into the fundamental CFT. This dimension arises from how quantum field theory organises itself across different energy scales. This is an example of dimensional emergence, where one spatial dimension is shown to emerge from quantum information. How about building AdS geometry from the entanglement of a CFT vacuum? Start with the CFT vacuum state $|\Omega\rangle$. This state has quantum fluctuations that create correlations across the field, making it entangled. For each boundary region A, calculate S_A. Now you have a function that maps regions to their entropies. This function alone allows you to reconstruct the bulk AdS geometry.

Step 1. The RT formula gives us S_A = Area(γ_A)/(4G_N), which means that for each region A, we have the area of its corresponding minimal surface.

Step 2. The metric determines these minimal surfaces. You constrain the metric by seeing how areas change when you vary the regions.

Step 3. Find the metric that reproduces all the minimal surface areas. The CFT vacuum distinctively determines the AdS metric for this purpose, although there is some ambiguity deep in the bulk.

The geometry is defined uniquely by the boundary entanglement, except for subtle gauge ambiguities. Researchers have clearly illustrated this in simple cases. We start with entanglement data, which is pure quantum information, and reconstruct the spacetime geometry from it. The metric emerges from the pattern of quantum correlations. Entanglement creates two things: flat space and curvature. In AdS/CFT, the bulk has constant negative curvature everywhere. This curvature arises from the conformal symmetry, which is the scale invariance of the CFT. For CFT states that break this symmetry, such as thermal states and excited states, the bulk exhibits varying curvature. This leads to the formation of black holes, the appearance of localised energy, and the propagation of gravitational waves, all resulting from boundary entanglement. Inject energy into the CFT, creating an excited state, and entanglement changes. The bulk geometry responds: its curvature varies, and black holes can form. The response in geometry comes from a reorganisation of entanglement. The RT formula makes this precise. Changing S_A for different regions results in the evolution of minimal surfaces. Each change is a reflection of a change in curvature. Entanglement not only creates distance but also the full structure of geometry, including curvature.

When it comes to building from one dimension to three, AdS/CFT shows how one dimension emerges: the radial direction, which comes from the CFT's energy scale structure. Here lies the clue, that it is possible for dimensions to emerge. But we need all three spatial dimensions to emerge. In AdS/CFT, the d-dimensional CFT produces a (d+1)-dimensional bulk. The extra dimension emerges from this; the original d dimensions are the same in both the boundary and the bulk. Or are they? The CFT itself might be emergent. If a (d-1)-dimensional quantum system, which has less spatial structure, gives rise to a d-dimensional CFT, you have demonstrated another layer of emergence. This is harder since no clean example like AdS/CFT exists. But tensor network models suggest it's possible. You can start with a (d-1)-dimensional lattice of qubits and entangle them in a suitable manner, which can lead to an emergent d-dimensional quantum field theory. Alternatively, *consider* recursive holography, where a d-dimensional CFT gives rise to a (d+1)-dimensional bulk. This bulk then becomes the boundary of a (d+2)-dimensional spacetime, and emergence stacks. Yet, there's a more direct approach, the one we will be pursuing in the coming chapters. Think about a quantum system with no pre-existing spatial dimensions, it only has quantum degrees of freedom, and these have no geometric structure and we will show how three-dimensional space emerges from its entanglement pattern. It involves using the lessons from AdS/CFT, entanglement creates geometry, the RT formula, minimal surfaces, and extending these concepts to show how realistic quantum systems produce three-dimensional emergent geometries. That covers Chapters 9-11.

The breakthrough is not a single formula, but the convergence of multiple concepts. The RT formula's exact match between entanglement and area, derived earlier in this chapter, converges with tensor network constructions that make geometry computable, and the error-correction framework that explains why holography works. A conjecture in Chapter 6 has become a general principle: geometry is entanglement, made precise. Why does three-dimensional space emerge rather than two or four? You can find the answer in Chapter 10. But first, in Chapter 9, we will see explicit constructions, including how to build space from qubits, how entanglement patterns shape geometry, and how to move from abstract quantum information to tangible emergent spacetime.

9

Building Space from Qubits

Every structure needs a blueprint. For a building, that means floor plans, load-bearing walls, and electrical routing. For spacetime emerging from quantum information, a tensor network is the blueprint. A tensor network represents a graph that shows how quantum systems connect. Nodes in the graph represent qubits, which are quantum two-level systems. The edges in the graph represent the entanglement between these qubits. The topology, which determines how the qubits link to each other, decides everything: the amount of entanglement that exists, how it distributes, and the geometry that emerges. Change the topology, and you change the geometry. For instance, a linear chain of qubits produces one-dimensional emergent space, while a square lattice produces two dimensions. Meanwhile, a hierarchical tree structure generates hyperbolic geometry, which includes an extra radial dimension, the same as that found in anti-de Sitter space. To get the emergent metric, you specify the network, calculate the entanglement entropies, and it falls out. In 2019, researchers built one of these networks that is a simplified version on a quantum processor. They measured entanglement across different regions and found that the Ryu-Takayanagi formula held: quantum correlation equalled geometric area in the emergent space. This shows the blueprint works. Now we need to find out which blueprint produces three spatial dimensions.

Qubits are the simplest quantum systems, which are two-dimensional

Hilbert spaces. They are the quantum versions of classical bits, but with a defining difference: bits are either 0 or 1, whereas qubits exist in superposition, represented as $\alpha|0\rangle + \beta|1\rangle$. Why build space from qubits rather than continuous quantum fields? Several reasons make this the better starting point. *Firstly*, qubits are simpler because they are discrete and countable, making them easier to visualise and calculate with than quantum fields, which have infinitely many degrees of freedom. *Secondly*, qubits are universal according to quantum computation theory, any quantum system can be approximated using qubits, so if space emerges from quantum information, it should be possible to emerge from qubits. *Thirdly*, qubits are practical. We have the technology to build physical qubits in laboratories, such as superconducting circuits, trapped ions, and quantum dots, which raises the possibility of creating emergent spacetime in a lab if geometry emerges from entangled qubits.

Qubits are the natural starting point. If we can show that three-dimensional space emerges from entangled qubits, we will have demonstrated how dimensionality arises from the quantum information structure. To build space from qubits, we require mathematical tools for describing entanglement, which are provided by tensor networks. A tensor is an array with multiple dimensions, made up of numbers. It can be a vector, which is a tensor of rank-1, or a matrix, which is rank-2. Tensors with higher ranks have more indices. In quantum mechanics, a tensor is used to represent quantum states that exist across multiple subsystems. When we have two qubits in a state $|\psi\rangle$, we can write:

$$|\psi\rangle = \Sigma_{i,j=0}^{1} T_{ij} |i\rangle|j\rangle$$

Here, T is a 2×2 matrix, which is a rank-2 tensor. The entries T_{ij} are complex numbers that specify the amplitude for each basis state. Scale up to N qubits, and you will need a rank-N tensor with N indices, each running from 0 to 1. This results in 2^N complex numbers, which are exponentially large. But most physical states do not require all that information. They have structure, including locality, limited entanglement, and specific correlation patterns. Tensor networks exploit this structure, which enables efficient representation of states. We have already seen these components in Chapter 7, which include nodes carrying local quantum information, edges representing entanglement bonds, and contraction, where summing over shared indices gives us the full quantum state.

The network topology, which is how nodes connect, encodes the entanglement structure. This results in different topologies producing different states, different patterns of entanglement, and different geometries that emerge. For example, the GHZ state. *Consider* the three-qubit Greenberger-Horne-Zeilinger state, often referred to as the GHZ state.

$$|GHZ\rangle = (|000\rangle + |111\rangle)/\sqrt{2}$$

All three qubits exhibit maximal entanglement. If you measure one and find $|0\rangle$, the others must be $|0\rangle$. If you find $|1\rangle$, the others are $|1\rangle$. We can represent this as a tensor network.

Imagine a star with three outer nodes that are connected to a central vertex. This topology precisely encodes the entanglement structure. The maximal three-way entanglement of the GHZ state requires all three qubits to describe the correlations. Any pair of qubits, considered alone, is unentangled. If you trace out one qubit, the remaining two form a product state. Different topologies create different entanglements. A chain, where qubit 1 is connected to 2 and then 2 to 3, produces different correlations than the star configuration. The topology determines the physics. Consider a one-dimensional quantum system consisting of N qubits arranged in a line.

1—2—3—4—...—N

Each qubit, i, has a physical index, p_i, which can be either 0 or 1. The quantum state is:

$$|\psi\rangle = \Sigma\_\{p_1,\ldots,p\_N\}\ \mathrm{Tr}(A\^\{p_1\}\ A\^\{p_2\} \ldots A\^\{p\_N\})\ |p_1p_2\ldots p\_N\rangle$$

Each A^{p_i} is a matrix with a dimension of $\chi \times \chi$, known as the *bond dimension*. This is a tensor network where each site contains a rank-3 tensor: it has one physical index (the qubit state) and two virtual indices (which connect to its neighbours). The bond dimension χ decides the amount of entanglement the state can have. Larger χ means more entanglement. If χ is 1, the state has no entanglement and is a product state. When χ equals 2^{N/2}, the state can be any state, but you've lost efficiency because there are exponentially many parameters. For ground states of local Hamiltonians, where qubits mainly interact with their neighbours, modest bond dimensions are sufficient ($\chi \sim 10$-100). This is because of area-law entanglement. The entanglement entropy between a region and its complement scales with the boundary; in one dimension, it is just the two endpoints. Matrix Product States (MPS) make use of this structure, which allows them to represent one-dimensional quantum states efficiently with polynomial resources instead of exponential ones.

What is meant by one-dimensional emergent space in the context of an MPS? The MPS has linear topology, with qubits arranged in a line, which indicates a single spatial dimension. However, there's a subtlety: the MPS describes qubits on a lattice, a discrete one-dimensional structure. Is this structure fundamental or emergent? It depends on your perspective. If you start by lining up qubits, they impose the one-dimensionality. However, if you start with abstract qubits and ask, "What network topology minimises energy for a given Hamiltonian?" the linear arrangement emerges as a natural occurrence. For Hamiltonians with nearest-neighbour interactions, the ground state forms an MPS with a linear topology. The structure emerges from the dynamics, rather than from an imposed geometry. This suggests emergent dimensionality. The spatial structure which qubits are "close" to each other emerges from entanglement minimisation, rather than from a pre-existing space. For two-dimensional quantum systems, Projected Entangled Pair States (PEPS) are a generalisation of MPS. Instead of a line, qubits sit on a 2D lattice, a square grid, for example. Each qubit has a rank-5 tensor, which includes one physical index for the qubit state and four virtual indices connecting to its four neighbours: up, down, left, and right. The PEPS network topology consists of two dimensions. This raises the question: does two-dimensional space emerge? The answer is yes, in the same sense as MPS. If you start with qubits and ask what network topology minimises energy for a 2D local Hamiltonian, you will arrive at a PEPS on a square lattice. The two-dimensional geometry emerges from the entanglement pattern. Qubits in close proximity within emergent space demonstrate significant entanglement. Those that are far apart have weak entanglement. The network topology determines the spatial structure. PEPS can describe rich physics, such as 2D quantum spin systems, topological phases, and fractional quantum Hall states. The emergent space features properties such as curvature if the lattice curves, or topology if it contains holes or handles. However, PEPS only gives two dimensions - there are no radial directions. To get three emergent spatial dimensions, we need something more.

The Trouble

Is it possible to create a tensor network with three spatial dimensions? Yes, but it is not just a matter of placing qubits on a 3D lattice. You can build a 3D tensor network where each qubit has seven indices: one physical and six virtual, connecting to neighbours in three-dimensional space. This network, when contracted, gives a quantum state with a 3D spatial structure. But this approach has its problems. Contracting a 3D tensor network is computationally expensive, to the point of being intractable for large systems. It's really helpful to get a clear understanding of this. The computational cost of contracting a tensor network is related to our mathematical methods, not to any limitation of nature. Nature simply evolves, without contracting tensor networks. When we refer to higher-dimensional networks as 'intractable', we are highlighting a symptom, rather than the underlying reason. The physical reason why certain geometries do not occur is that the required entanglement structure is not feasible. The *monogamy of entanglement* means that as qubits are connected to more neighbours in higher dimensions, their correlation capacity is reduced, and this decreases below the threshold needed to maintain long-range geometry. The computational difficulty we experience is a reflection of this physical reality, rather than its cause. This distinction has several implications — one we will return to in Chapters 10 and 11. Worse, this forces a three-dimensional structure from the beginning. It assumes that three dimensions exist and builds a network to fit this assumption. We want the opposite: to start with quantum information, without assuming any dimensionality, and demonstrate how three-dimensional geometry emerges.

The Hierarchical Answer

The solution lies in hierarchical networks, extended to have three spatial dimensions rather than two plus one. Standard MERA with a 1D boundary, which is a line of qubits, produces a 1D boundary plus a 1D radial, resulting in a 2D emergent space. Generalised MERA with a 2D boundary, which is a plane of qubits, and a hierarchical structure produces a 3D emergent space that includes a 2D boundary plus a 1D radial dimension. This has been explored in the literature. The Generalised MERA is complex, as each layer consists of a 2D PEPS that connects through isometries and disentanglers across layers. Calculations show that this structure produces a 3D emergent geometry with negative curvature. The entanglement structure satisfies the 3D area laws: S_A ~ (the 2D boundary area of A). Three emergent spatial dimensions are achievable through tensor networks, but these constructions start with a 2D boundary. This explains the emergence of the radial direction, but it does not clarify why the boundary has two dimensions instead of one or three. Why Two Dimensions on the Boundary? We are now approaching some fundamental questions. In AdS/CFT, the boundary has d dimensions, while the bulk has d+1 dimensions. One dimension emerges through the RG flow, which is the energy scale structure. But why is the boundary d-dimensional? Is this a fundamental property or an emergent one? In standard AdS/CFT, you choose the boundary dimensionality - you can opt for a 2D CFT, a 3D CFT, or a 4D CFT. You might start without assuming any dimensionality, focusing solely on quantum degrees of freedom and qubits, along with a specific Hamiltonian. The question is: what kind of dimensionality emerges when we minimise energy, maximise entanglement stability, or satisfy other physical principles? This is more difficult. Unlike AdS/CFT, we do not have a straightforward example. Studies on tensor networks show that dimensionality is a property that comes from the system itself. It arises from how several key features interact with each other. The interaction between qubits and their nearest neighbours naturally imbues dimensionality into the interaction graph, effectively seeding it with geometry through locality. Furthermore, the monogamy of entanglement plays an integral part in shaping the network topology by limiting the extent of mutual entanglement among qubits, thereby constraining the possible configurations. The stability of these topologies under perturbations is a critical factor, as some configurations endure while others dissolve, and it is the stable ones that seem to favour specific dimensionalities, as if the network gravitates towards geometrically coherent structures. Beyond this, networks that optimise entanglement entropy for a given correlation structure tend to settle into configurations with preferred dimensionalities, implying that dimensionality may, to some extent, be an emergent property of an optimisation process, where geometry arises as the most efficient solution for organising quantum information.

The hypothesis I'll develop in Chapter 11 is that these constraints provide a framework where 3D might emerge when certain constraints are imposed to produce a 3D emergent space from a 2D boundary plus 1D radial. Not because three is special, but because it is the number that works; it remains stable, is optimal, and arises from quantum information under realistic constraints. Another approach is graph states. A graph state is defined by a graph, which is a network of vertices and edges, where each vertex represents a qubit. The state is constructed by, Starting with all qubits in $|+\rangle = (|0\rangle + |1\rangle)/\sqrt{2}$ and then to apply a controlled-Z gate between qubits i and j for each edge (i,j). The resulting state is highly entangled, with its structure determined by the graph. Different graphs produce different entanglement patterns such as, a line graph has a 1D entanglement pattern, a square lattice has a 2D pattern, a cubic lattice has a 3D pattern and a tree graph has a hierarchical pattern with an emergent radial dimension. For graph states, one can calculate entanglement entropy using analytical methods. For region A, S_A is roughly equal to the number of edges crossing ∂A. This follows an area law, where entropy scales with the boundary, not the volume. Graph states show emergent geometry. The graph defines the distance between qubits as the shortest path length, which creates a discrete metric.

For regular lattices, like square and cubic ones, the emergent metric is flat. For tree graphs, the metric has negative curvature which is a hyperbolic space. The curvature varies with different graphs. Graph states show that the entanglement structure in a graph determines its emergent spatial structure. If you change the graph, you change the geometry too. Tree graphs hold a lot of importance. A tree is a graph with no loops, where each vertex connects to a root in a unique way. In a regular tree, all vertices except the leaves have the same number of children. In a tree where the branching factor is b, the number of vertices at a distance r is: $N(r) \sim b^r$

This grows at an increasing rate with distance, which is the signature of hyperbolic space, characterised by negative curvature. In flat space, $N(r) \sim r^{d-1}$ (polynomial growth). In hyperbolic space, $N(r) \sim e^{(d-1)r}$ (exponential growth). A tree graph with exponential growth corresponds to hyperbolic geometry. The space that emerges has negative curvature, like AdS. This explains why hierarchical networks with tree-like structures produce AdS-like geometries. The hierarchical network generates a hyperbolic space as a natural outcome. For flat or positively curved emergent space, various network topologies require graphs characterised by polynomial or sub-polynomial growth. Let us return to quantum error correction, which provides another perspective on emergent geometry. A quantum error-correcting code encodes logical qubits into multiple physical qubits. The code subspace, which comprises the allowed states for physical qubits, has a structure that protects against errors. Recent work shows that holographic codes, which are codes with a geometric structure, can produce emergent dimensions. This scales up. A 3D holographic code has a 2D boundary, which is a sphere of physical qubits, and a 3D bulk, which is a ball of tensors. Two spatial dimensions in the bulk emerge from the structure and topology of the 2D boundary. But this gives only two emergent dimensions if your boundary is 2D. To obtain three spatial dimensions, you need one of the following: (A) A 3D boundary provides three spatial dimensions directly. (B) Recursive holography has a 2D boundary that is holographic from a 1D sub-boundary. (C) A different structure exists where dimensionality is not just the boundary dimension plus one. Option (B) is intriguing: nested layers of holography. A 0D system, with few qubits, produces a 1D boundary. This 1D boundary, in turn, holographically produces a 2D bulk. The 2D bulk then produces a 3D space. This results in three spatial dimensions that emerge in a recursive manner. It suggests, though, that dimensionality could emerge from multiple layers of holography, determined by the entanglement structure and QEC properties.

You may ponder is it possible to create emergent spacetime in a lab environment? Not yet, but things are picking up pace now. Quantum computers have now reached 50-100 qubits. Researchers are preparing states, measuring entanglement entropies, and testing area-law scaling. Several experiments have created states similar to graph states or small tensor networks, which verify that the entanglement structure determines the emergent correlation length and effective dimensionality. In 2019, researchers implemented a simplified code on a quantum processor. They measured entanglement entropies for various boundary regions, which confirmed the RT formula held, showing that quantum entanglement was equal to the size of minimal network cuts. This represented emergent geometry on a chip, with a discrete network of qubits rather than continuous spacetime, and the principle still applies. Future quantum computers, having thousands or millions of qubits, will be able to implement more complex tensor networks that approach continuous geometry on a large scale. Perhaps within a few decades, we might be able to create a quantum system in which genuine 3D spacetime emerges from entangled qubits. This could allow us to probe the emergent geometry, create "bulk particles," and test how gravity emerges. It would be the ultimate demonstration of creating space in a laboratory. All the tensor networks we have discussed are discrete and have finite qubits and discrete connections. Yet the universe appears to be continuous. How can we bridge this gap? Increasing the number of qubits and decreasing the lattice spacing reaches the continuum limit, which causes the discrete structure to transition to a continuous one. For instance, a digital photograph. At low resolution, individual pixels are visible. At high resolution, the image appears smooth and continuous. The continuum limit works similarly: with enough properly spaced qubits, it approximates continuous space. To make this rigorous from a mathematical standpoint is a delicate task. You need to carefully adjust the bond dimensions, the structure of entanglement, and the topology of the network as you refine the lattice. For ground states of local Hamiltonians in d dimensions, the continuum limit is well-defined. The discrete tensor networks, like MPS and PEPS, approach quantum field theory with the emergent geometry becoming continuous. For holographic networks, such as hierarchical structures or generalised networks, the continuum limit produces geometries that are AdS or AdS-like. The insight is that a discrete network can approximate curved spacetime to any desired level of accuracy. With a sufficient number of qubits, the discreteness becomes undetectable. Building space from qubits does not imply that spacetime is discrete. Spacetime emerges from quantum information, which we can describe either discretely with qubits or continuously with quantum fields, based on our choice of description.

One question remains: why three?

10

Why Three Dimensions?

What is the reason for having three spatial dimensions? We have shown that space emerges from quantum entanglement. The AdS/CFT demonstrates one dimension emerges from the structure of energy scales. Tensor networks show that specific entanglement patterns give rise to specific geometries. A generalised MERA with a 2D boundary is capable of producing a three-dimensional space with an emergent 3D geometry. But what makes the universe pick three over two or four? The answer isn't anthropic as Chapter 2 argued. We exist because three dimensions allow the complexity needed for life. This still doesn't explain why the universe has three dimensions in the first place; it only explains why we see them that way. The answer might be optimisation. Quantum information, constrained by the principles of quantum mechanics, forms a three-dimensional structure. Not because three is special, but because it works; it's stable, it's optimal, and it's what emerges when quantum entanglement networks meet realistic physical requirements. But again, this answer is frustrating. The optimisation argument had me frustrated, as it appeared circular: three dimensions are optimal because the constraints select three, but who chose these constraints? However, it eventually became clear that the constraints are not chosen but rather they are a direct result of quantum mechanics itself. Monogamy, area-law scaling, and locality are theorems, not assumptions. The argument, thus, is not circular, but rather it rests on a solid foundation, being derived from the principles of quantum mechanics.

Let us start with quantum information, which consists of abstract degrees of freedom that have no pre-existing spatial structure. This involves a Hilbert space, a Hamiltonian that governs dynamics, and quantum entanglement that creates correlations. What constraints govern how this system is organised? The constraints are not imposed arbitrarily, but are instead inherent to quantum systems, which are governed by local interactions and reflect the fundamental principles of quantum mechanics and information theory. They also decide which geometries are likely to emerge.

Consider the reasons why two spatial dimensions are insufficient. Take a 2D tensor network: qubits on a square lattice, where each qubit is entangled with its four nearest neighbours (±x, ±y). The entanglement entropy of region A, with a boundary of length L (measured in bonds cut), is bounded by:

$$S\_A \leq L \cdot \log \chi$$

This is the area law: entropy scales with boundary length, not area. The prefactor log χ reflects the entanglement capacity per bond. Does this 2D network have the capability to support complex hierarchical structures?

Problem 1. Insufficient Entanglement Capacity. In 2D, a region with radius R has approximately R^2 qubits, denoted as N. The length of the boundary is approximately R. The entanglement entropy scales as S, which is proportional to R. Therefore:

$$S/N \sim R/R^2 = 1/R$$

As regions grow, this ratio decreases. The entanglement per qubit decreases as the regions grow larger. The system is unable to maintain long-range correlations, which limit its information processing capacity.

Problem 2. Limited Connectivity. Each qubit connects to only four nearest neighbours. This limits the number of independent paths between distant qubits. The network is "sparse" because it lacks enough connectivity for complex correlation patterns. In information-theoretic terms, this limits the channel capacity. The flow of quantum information through the network depends on the number of independent paths that are available.

Problem 3. Topological Constraints. In 2D, topological defects are like points. Since knots and links need three dimensions to be nontrivial, they cannot exist. This limits the types of quantum phases and topological structures possible. Many robust quantum error-correcting codes rely on 3D topology, such as the 3D toric code. These structures do not exist in 2D, which limits the system's ability to protect quantum information from errors.

Problem 4. Fragility Under Perturbations. 2D networks with area-law entanglement are unstable. Small perturbations, such as adding interactions or making minor adjustments to bond dimensions, can lead to changes in the entanglement structure. The system lacks a sufficient "buffer" to maintain its coherence. Numerical simulations confirm this. For realistic Hamiltonians, 2D PEPS must have bond dimensions χ of around 10-50 to capture ground states. However, changing a Hamiltonian parameter by 1% can double the required χ. This shows that the states are fragile.

Why Not Two?

Now *consider* adding a hierarchical structure, creating a 2D version with a 1D boundary (a line of qubits) and extra layers, resulting in a 2D emergent space (1D boundary + 1D radial). The same issues arise again. Compared to a 2D boundary, a 1D boundary have limited resources, and the area law results in $S \sim 1$, reflecting the simplicity of just two boundary points. This simplicity means the hierarchical structure cannot introduce enough complexity. The hierarchical structure cannot add enough complexity. Two dimensions, whether from 2D networks or 1D + 1D radial, are too restrictive. They have insufficient entanglement capacity, limited connectivity, and an inadequate topological structure. The emergent geometry that results is fragile and cannot support the rich physics that we observe in nature.

Why Not Four?

Imagine there are four spatial dimensions. *Consider* a 4D tensor network: qubits are arranged on a 4D hypercubical lattice, with each having eight nearest neighbours (±x, ±y, ±z, ±w). For region A, which has a 3D boundary surface area of A∂, the area law gives:

$$S\_A \sim A\_\partial \cdot \log \chi$$

The boundary is now a 3D volume, entropy scales with this 3D "area." This seems promising. Having more neighbours means there are more entanglement capacities. But problems emerge.

Problem 1 Diluted Entanglement. In 4D, each qubit shares its entanglement with eight neighbouring qubits. The monogamy principle applies, which means that the total entanglement each qubit has is limited. With eight bonds instead of six (3D) or four (2D), each individual bond carries less entanglement. If the total entanglement per qubit is roughly log χ, the strength of each bond is:

2D: (log χ)/4 per bond, 3D: (log χ)/6 per bond, 4D: (log χ)/8 per bond. To maintain the same bond strength in 4D as in 3D, we require the following condition:

$$\chi\_4D \sim \chi\_3D\hat{}(4/3)$$

This condition arises from a physical constraint, rather than a mere calculational limitation. Quantum mechanics imposes a strict limit on the amount of entanglement a single qubit can share, known as the monogamy of entanglement, which means that spreading correlation across eight neighbours instead of six results in each bond being physically weaker. As a result, the geometry that relies on these bonds to maintain coherence loses its structural integrity. The increased χ required is a direct consequence of this physical limitation, rather than a cost solely imposed by our simulation methods.

Problem 2 Physical Instability from Entanglement Dilution. In four spatial dimensions, each qubit's entanglement budget must be divided among eight neighbours, rather than six. The monogamy of entanglement, a fundamental theorem in quantum mechanics proven by Coffman, Kundu, and Wootters in 2000, sets a strict limit on the total correlation any qubit can maintain. When this budget is divided eight ways instead of six, the long-range correlations that support an emergent geometry decay more rapidly with distance. If the bond strength falls below a critical threshold, the geometric structure loses coherence completely: regions that should be close in the emergent space become effectively decorrelated, causing the geometry to dissolve. Theoretical analysis confirms that this situation represents a physical failure, rather than just a computational one, which clearly indicates that entanglement networks are inherently unstable in higher dimensions, although researchers are still working to prove this across all possible quantum systems.

The cost of contracting a 4D tensor network scales sharply, at roughly χ^8 per site, which is steeper than the χ^6 scaling for 3D networks. This mirrors the physical reality that the entanglement structure in four dimensions is more fragile and requires higher bond dimensions to maintain coherence. Physicists have a clear reason to disfavour four dimensions, and it is the physical instability of these systems, not the difficulty of simulating them.

However, the issue runs deeper than just raw bond dilution. Approximate contraction of tensor networks, the method by which physical predictions are extracted, reflects a qualitative phase transition between d=3 and d=4. Contracting a 2D network uses 1D boundary methods (matrix product states), which are efficient. Contracting a 3D network uses 2D boundary methods, which are hard but feasible. Contracting a 4D network requires 3D boundary methods, which themselves require 2D boundary methods; the difficulty compounds. Each additional dimension nests another layer of complexity. This cascading structure reflects a fundamental physical truth: in four dimensions, the entanglement hierarchy needed to maintain coherent long-range geometry cannot be sustained under realistic physical constraints. The contraction difficulty reveals the physical problem, rather than causing it.

Problem 3 Unstable Emergent Geometry. Higher dimensions allow more ways for perturbations to spread. A 4D network has many paths for errors to propagate. Quantum error correction becomes much harder and you have to protect against errors in more dimensions. While the 4D toric code is notable for having a macroscopic energy barrier against errors (making it a self-correcting quantum memory), it achieves this at the cost of enormous overhead: the number of physical qubits required per logical qubit grows much faster than in 3D, making it impractical for large systems. The trade-off between passive error protection and resource overhead becomes increasingly unfavourable in higher dimensions.

Problem 4 Overconstrained Error Correction. Holographic codes have a bulk that emerges from the code subspace. The code structure becomes problematic when a 4D bulk is derived from a 3D boundary because the 3D boundary tensor network is overconstrained physically. The entanglement structure of the 3D boundary is insufficient to encode a coherent 4D bulk due to the monogamy constraints on the 3D boundary qubits. As a result, the holographic encoding of bulk operators to boundary operators is not feasible, and encoding and decoding operations cannot be performed. This limitation arises from the physical feasibility of the code, which exists mathematically but lacks the necessary entanglement resources to operate effectively.

Problem 5 Inherent Geometric Instability. The emergent geometry in 4D possesses an unstable nature. Trying to create 4D emergent spacetime with matter or energy, which corresponds to excited states of the entanglement network, produces instabilities. The perturbations grow rather than decay. This means stable, long-lived structures are unable to form. Four dimensions are overwhelmed because they have too many neighbours, which hinders the effective distribution of entanglement, as the numerous paths for error propagation make the entanglement structure physically incoherent. As a result, the emergent geometry is unstable and cannot support robust error correction. In principle, you can construct 4D tensor networks, but they are not favourable, not optimal, and not stable. This result has a visceral quality to it. Four-dimensional space is not just difficult to construct, it actually falls apart. The entanglement network attempts to sustain geometry across a multitude of connections, but the monogamy of entanglement says it cannot be done. Each qubit has limited correlations to spread around, and when it tries to connect with too many others, the bonds weaken and the structure collapses. Nature has a similar limitation, which is why it does not build in four dimensions, much like how it is nearly impossible to have deep friendships with a thousand people: the available resource is limited, and spreading it too widely destroys its value.

Why do three dimensions work? Visualise a 3D tensor network: qubits are arranged on a cubic lattice, each having six nearest neighbours (±x, ±y, ±z). For region A with a 2D boundary surface area A∂:

$$S\_A \sim A\_\partial \cdot \log \chi$$

Why is three the optimal number? One of the best answers is balanced connectivity. Six neighbours provide the right balance. There are sufficient connections for rich correlation patterns without overwhelming entanglement monogamy. Each bond can be strong and it is not diluted; as in 4D, additionally it offers more pathways than the four neighbours in 2D.

A three-dimensional region of radius R contains roughly R^3 qubits, and its boundary area scales as R^2. The entanglement-to-qubit ratio:

$$S/N \sim (\log \chi) \cdot R^2/R^3 = (\log \chi)/R$$

The decrease in this is gentle, as expected from the area law, and it happens as the region size increases. The ratio S/N ~ 1/R applies in all dimensions (since S ~ R^(d-1) and N ~ R^d for any d), but the total information capacity and the physical cost of maintaining coherent entanglement both change with dimension. For a fixed number of qubits N, the entanglement scales as S ~ N^((d-1)/d), so:

- d=2: S ~ N^(1/2), growing slowly with system size
- d=3: S ~ N^(2/3), growing fast enough for complex physics
- d=4: S ~ N^(3/4), growing fastest but the monogamy constraint limits the entanglement, forcing it to be divided among eight neighbours per qubit, which in turn dilutes the strength of each bond, falling below the threshold necessary for a stable long-range geometry, and this physical instability is exacerbated by the cascading contraction problem, rendering the situation untenable under realistic conditions.

The 2/3 exponent in 3D is significant. It is large enough that large regions retain substantial entanglement capacity, supporting long-range correlations. However, distant regions decouple enough to allow effective field theory descriptions, meaning we can treat regions independently at low energies. Moreover, topological richness. Three dimensions support non-trivial knots, links, and braiding. These are impossible in 2D, as everything unknots, and they become trivial in 4D and higher, where knots can be untied using the extra dimension. 3D is the only one that provides the right topological structure for complex quantum phases and robust error correction codes.

In three dimensions the entanglement structure is self-sustaining, a direct consequence of physical tractability. With six neighbours per qubit, each bond carries sufficient correlation strength, avoiding the issues of dilution found in 4D and sparsity in 2D, thereby maintaining coherent long-range geometry. The monogamy budget is well-distributed, allowing each qubit to share strong correlations with its neighbours, which in turn enables the emergent geometry to remain stable under realistic perturbations. Researchers have developed tensor renormalisation group methods over the past two decades, effectively approximating the physical ground states of 3D systems with area-law entanglement. Bond dimensions of χ in the range of 20-100 are physically sufficient, reflecting the genuine stability of the underlying entanglement structure, rather than just being computationally tractable. Crucially, three dimensions are not just simpler to simulate than four; they are physically distinct, with coherent entanglement networks that are not found in four-dimensional systems. Quantum error-correcting codes in 3D, such as the 3D toric code, have optimal properties. They strike a balance between code rate and distance, enabling efficient error correction with a reasonable overhead. In the holographic picture, a 2D boundary, which is a sphere, encodes a 3D bulk and produces a code with strong parameters. The Ryu-Takayanagi formula tells us that the recoverable information from a boundary subregion of size r is proportional to the minimal surface area ~ r^(d-2), while the total boundary qubits in that subregion scale as r^(d-1). The local code rate is therefore:

$$k/n \sim r^{(d-2)}/r^{(d-1)} = 1/r$$

This decreases with subregion size, indicating redundancy in the encoding, a hallmark of good error correction. The code distance—a measure of how many physical qubits must be corrupted before logical information is lost, is proportional to the system size. The code is therefore resilient. The qualitative difference between dimensions is evident in the overall structure. For a 3D bulk from a 2D boundary, the holographic code has a manageable structure: the entanglement in the 2D boundary is physically sufficient for the encoding. For a 4D bulk from a 3D boundary, the entanglement structure of the 3D boundary is physically overconstrained. This is because the correlations needed to encode a coherent 4D bulk are more than what monogamy allows on a 3D boundary, which means the code cannot be efficiently decoded. A code that cannot be decoded is useless, no matter what its formal parameters are. Hierarchical Structure, for holographic emergence, where 2D boundary and 1D radial combine to form 3D, the 2D boundary has sufficient capacity. A 2D lattice of qubits is capable of encoding complex quantum states. In contrast to a 1D boundary, which produces a 2D bulk, there is insufficient capacity. A 3D boundary, which produces a 4D bulk, faces excessive constraints. Three dimensions are optimal in terms of all metrics, including connectivity, capacity, topology, physical stability, error correction, and hierarchical structure.

Think about the tensor network as a graph, with qubits as vertices and entanglement bonds as edges. The graph's properties are connectivity, diameter, clustering, and growth rate, and now determine its emergent geometry. The key graph-theoretic properties change in a systematic way with dimension. Each vertex has 2d neighbours, which means 4 neighbours in 2D, 6 in 3D, and 8 in 4D. The maximum distance between vertices in a region of size N increases as $N^{(1/d)}$. The clustering coefficient, which shows how often a vertex's neighbours connect to each other, decreases as the dimension increases, whereas the growth rate, which is the number of vertices within distance r, increases as r^d.

These properties decide how entanglement spreads, the flow of information, and the complexity of the quantum states. Graph-theoretic analysis, which involves examining information capacity, average distance, clustering, and connectivity across dimensions, points to d=3 as a balanced choice.

The analysis across dimensions reveals interesting insights. One dimension proves inadequate with only two neighbours and an average distance that scales linearly with the number of nodes, N. In contrast, two dimensions offer some improvement with four neighbours, and the average distance scales roughly as N^(1/2), yet the connectivity still falls short for complex patterns. However, three dimensions strike a balance, with an average distance of approximately N^(1/3), and six neighbours provide sufficient connectivity along with reasonable clustering. Beyond three dimensions, the networks become overly connected, which is not desirable. In four dimensions, the average distance decreases to N^(1/4), making it shorter, but this comes with a drawback. The monogamy of entanglement forces each of the eight bonds to carry less entanglement, thereby compromising the long-range correlations that are essential for sustaining emergent geometry. The physical cost of maintaining coherence increases as a function of N^(4/3) per qubit, which makes it prohibitively expensive.

If entanglement networks are analysed for information flow, stability, and physical coherence, then graph theory suggests d=3 is the best choice. *Take* a quantum system with N qubits on a d-dimensional lattice. The total dimension of the Hilbert space is 2^N. However, most states are not physically accessible because they violate locality, have excessive entanglement, or cannot be ground states of local Hamiltonians. The physical subspace dimension can be estimated from area-law scaling. A region's entanglement entropy is roughly equal to its boundary area, which is proportional to N^((d-1)/d). For a pure state of the full system, this entropy determines the Schmidt rank, the number of terms needed in the Schmidt decomposition across any bipartition. The dimension of physically accessible states scales as:

dim(physical) ~ e^(S_typical) ~ e^(C N^((d-1)/d))

where C is a constant that depends on the bond dimension and interaction strength.

Comparing across dimensions: In one dimension, the effective state space is a constant, e^C, independent of system size. The system cannot encode growing information. In two dimensions, it grows as e^(C√N): real growth, but sluggish. Three dimensions accelerate this to e^(CN^(2/3)). Four dimensions push further to e^(CN^(3/4)), the fastest growth rate. More capacity sounds better, but it comes at a cost.

Higher dimensions allow for larger physical subspaces, providing more room for information. But there's a trade-off. Higher dimensions also bring,

Higher dimensions impose entanglement monogamy pressure which weakens each bond as correlations are distributed across a larger number of neighbours, thereby increasing physical fragility. The presence of more dimensions also gives rise to a greater number of error channels and tighter constraints due to monogamy, resulting in entanglement being spread more widely across neighbours. The optimal dimension balances information capacity against physical stability and geometric coherence. Information-theoretic analysis suggests d=3 provides the best balance. Below d=3, capacity is too limited. Above d=3, entanglement dilution and geometric instability become dominant. A quantum error-correcting code encodes k logical qubits into n physical qubits, with a distance d_code, which is the minimum number of errors needed to corrupt the logical information. The code parameters satisfy the quantum Singleton bound:

$$k \leq n - 2d\_code + 2$$

Good codes approach this bound by maximising k (information capacity) and maintaining a large d_code (error protection). In holographic codes, which are geometrically structured codes where bulk emerges, the bulk dimensionality has a close connection to the code parameters. Recent work has shown that:

- 2D bulk (from 1D boundary). The code rate k/n is low, which means it has a limited information capacity. The code is inefficient as it requires many physical qubits per logical qubit, resulting in poor error protection.

- 3D bulk (from 2D boundary). The code rate can be optimal. It achieves a good balance between k and d_code. This provides strong error correction with a reasonable overhead. The 3D toric code and related constructions are examples of this. Crucially, the 2D boundary entanglement structure is physically sufficient to support encoding, which means the code can be practically used.
- 4D bulk (from 3D boundary). Even when the formal code parameters are acceptable, the 3D boundary's entanglement structure is physically overconstrained; monogamy limits on the 3D boundary qubits mean the correlations needed for a coherent 4D bulk cannot be sustained. Consequently, the encoding and decoding operations needed for error correction cannot be performed in any physical realisation. The code parameters are theoretically sound, but they have no corresponding physical entanglement resources to support their operation.

In the holographic setting, the most efficient quantum error-correcting codes are achieved with three dimensions. This is not a coincidence. Emergent spacetime is about encoding quantum information efficiently and protecting it against errors, so dimensionality has to be linked to code optimisation. Is three the sweet spot? In quantum field theory, the renormalisation group, describes how physical theories change when you move across energy scales. The fixed points of RG flow are scale-invariant theories, which are known as conformal field theories. For a CFT in d boundary dimensions, AdS/CFT correspondence produces a (d+1)-dimensional bulk, where the radial direction in AdS corresponds to the RG flow direction. Why should the boundary be d-dimensional? The structure of RG flow gives us insight: The structure of RG flows changes greatly with dimension. In one dimension, flows are simple and have few fixed points, due to the c-theorem. The two-dimensional case is more complex, with many fixed points and a phase structure governed by the F-theorem. Three dimensions have the most complex behaviour: many fixed points, complex phase diagrams, and various phenomena, all constrained by the a-theorem.

However, beyond three dimensions, flows become simpler in some ways, in four dimensions and higher, perturbation theory often controls the flows, but non-perturbative effects make them complicated, and CFTs have fewer examples and less diversity.

The richest RG structure occurs at d=3, where we have the maximum number of fixed points, the most complex flows, and the greatest diversity. This corresponds to 3+1-dimensional spacetime. When nature chooses a boundary theory with optimal RG properties, such as a rich structure, stable fixed points, and maximal complexity, d=3 emerges naturally. Adding the emergent radial direction gives us 3+1 = 4 total dimensions, which include three spatial dimensions plus one time dimension. For a region of size R in d spatial dimensions, the volume scales as R^d while the boundary area scales as R^(d-1).

The entropy bound reveals that entropy scales as R^(d-1), not R^d.

- d=2: S ~ R, volume ~ R^2. Entropy seems "too small" relative to volume. The constraints limit the information capacity.
- d=3: S ~ R^2, volume ~ R^3. This balances. There is enough entropy for complexity without overwhelming the volume.
- d=4: S ~ R^3, volume ~ R^4. The entropy is large, but sustaining this entanglement physically demands more than what any realistic quantum system can provide, given the monogamy constraints and the need for stable long-range correlations that exceed the available entanglement budget.

The three dimensions draw a balance between information capacity (or entropy) and physical coherence (or volume). The ratio S/V ~ 1/R in 3D carries special properties: Local thermalization (small regions can equilibrate on their own), Effective field theory (UV and IR physics become separate) and Classical emergence (the large-R limit approaches classical behaviour). These features are essential for the universe we observe. They optimise in three dimensions. Stability is non-negotiable. A fragile emergent geometry that is easily disrupted by perturbations will not persist.

Consider a small perturbation to the Hamiltonian:

$$H \rightarrow H + \delta H$$

What happens to the ground state, and to the emergent geometry that comes with it?

d=2. Small perturbations can drastically change the entanglement structure. The system remains near criticality, making it susceptible to changes in topology. The ground states are fragile. Numerical evidence from 2D PEPS for spin models indicates that a 1% change in interaction strength can cause a doubling of the bond dimension to maintain accuracy. The entanglement structure remains unstable.

d=4. Small changes in many dimensions cause instability. With high connectivity, errors spread quickly. The geometry that emerges cannot stay coherent when faced with realistic disruptions. Numerical studies of 4D tensor networks confirm that the entanglement structure needs extensive fine-tuning to remain stable. The emergent geometry is highly sensitive, and small changes to parameters can cause it to lose coherence or require large increases in bond dimensions to maintain its basic structure.

d=3. The system is robust. Small perturbations lead to small changes in the entanglement structure. The emergent geometry remains stable. The network can absorb errors without collapsing. Tensor renormalisation group studies of 3D systems show stability across variations in parameters. Bond dimensions of around $\chi \sim$ 50-100 are adequate even when Hamiltonian parameters change by 10-20%. The geometry remains well-defined and persistent. Nature favours stability. If only stable geometries endure, then three dimensions are favoured. The theoretical arguments gain substantial credibility from numerical evidence, revealing an impressive consistency across various dimensions. Studies by Corboz and collaborators between 2010 and 2018 on PEPS tensors applied to Heisenberg and Hubbard models in two dimensions have found that ground states need bond dimensions of $\chi \sim$ 10–50 for accurate representation, but these states are fragile. Even small changes of 1–5% in interaction parameters demand substantially larger bond dimensions, which reveal the sensitivity of the entanglement structure to perturbations.

Three-dimensional systems exhibit a notably distinct behaviour. The tensor renormalisation group calculations for 3D Ising and Potts models, conducted by researchers such as Gu, Levin, and Wen between 2008 and 2015, have yielded consistent results: bond dimensions of $\chi \sim 20$–100 produce accurate outcomes that persist even when parameters are varied by 10–20%, indicating a well-defined emergent geometry. This discovery has substantial implications, highlighting a stark contrast with the properties of four-dimensional systems. The few studies that have attempted to use 4D tensor networks for simple spin models have found a major finding: bond dimensions need to exceed $\chi \sim 500$ to maintain coherent entanglement, which suggests a genuine physical instability in these four-dimensional networks, making them prone to losing geometric coherence without extreme fine-tuning, and rendering them computationally intractable for systems larger than around ten sites per dimension. The emergent geometry in these cases is poorly defined and dominated by physical instabilities, lacking a coherent structure. Machine learning provides an alternative perspective, shedding new light on the issue. The neural network quantum states, introduced by Carleo and Troyer in 2017 and built upon by Sharir and collaborators in 2020, demonstrate a telling preference for architecture: convolutional architectures that incorporate 3D spatial structures, using 3D kernels, consistently outperform their 2D and 4D counterparts in representing ground states of local Hamiltonians, as if the networks themselves are uncovering the fact that three dimensions are the natural realm of physical entanglement.

Numerical evidence confirms that three dimensions are the optimal choice, supporting the theoretical arguments.

Objections

Objection 1: "You have assumed area-law entanglement and locality. Yet, these assumptions may not be fundamental."

These properties are not assumptions, but rather observed characteristics of physical quantum systems. Every quantum field theory and lattice model studied has these features. Area law has been thoroughly established for one-dimensional gapped local Hamiltonians, as Hastings demonstrated in 2007, and numerical evidence provides strong support for its validity in higher dimensions; however, a general proof for dimensions beyond one remains an indispensable unsolved problem in mathematical physics. Giving up locality and area law means giving up on physical systems that are governed by local interactions. The question is why physical systems have these properties. The answer is that systems that do not include them are unstable, physically incoherent, or violate causality.

Objection 2: "String theory requires 10 or 11 dimensions, so how does this fit?"

String theory's extra dimensions are compactified, curled up at tiny scales, like the Planck length. What I'm explaining is why we experience three large spatial dimensions. Extra compact dimensions might exist, but they are either small and compactified, which renders them effectively invisible and irrelevant for the geometry we observe, or they emerge at different scales from various entanglement structures. This argument shows that the spatial structure we see at a large scale has three dimensions. The structure at the Planck scale could be different, but that does not contradict how 3D geometry emerges at larger scales.

Objection 3: "You're just restating the problem. Why have quantum systems got these particular constraints?"

The constraints follow from quantum mechanics itself. Monogamy is a proven theorem in quantum information theory, as shown by Coffman-Kundu-Wootters in 2000. Locality emerges from causality, which requires a finite information propagation speed for consistent dynamics. The area law follows from local interactions, as local interactions concentrate entanglement at boundaries. These are not arbitrary restrictions, but mathematical consequences of quantum mechanics and causality. The question of "why these constraints?" is asking "why quantum mechanics?", which is beyond the scope of this book. Given quantum mechanics, these constraints are mandatory.

Objection 4: "This feels circular. You assume that emergent spacetime has properties like ours and then argue that it must be 3D."

I'm not assuming spacetime properties. I'm deriving them from quantum information theory. The constraints, such as locality, monogamy, area law, stability, and physical coherence, are properties of quantum systems, not spacetime. I begin with quantum information governed by these constraints and show that three-dimensional geometry emerges as the optimal solution. Spacetime properties are the outcome, not the starting point. The concern about circularity would be valid if I assumed that "space has stable orbits, which is why it's 3D." However, I argue that "quantum systems with these information-theoretic constraints optimise to 3D, and it so happens that 3D has stable orbits." The geometry follows from the information structure, not the other way around.

Objection 5: "How can we test this experimentally?"

We cannot change the dimensionality of the universe, but we can simulate emergent geometries in quantum systems. Experiments with larger quantum computers, comprising hundreds to thousands of qubits, could be used to test whether 3D emergent geometries are more stable and robust than 2D or simulated 4D systems. We have the capability to prepare networks with different dimensionalities, measure their stability under perturbations, test their error correction properties, and determine which dimension provides optimal performance. This would serve as a direct test of the optimisation arguments presented here.

Why three might be Inevitable. If this framework is correct, three dimensions emerge as the natural solution because quantum entanglement networks must simultaneously satisfy locality of interactions, the monogamy of entanglement, area-law scaling, physical stability of emergent geometry, stability against perturbations, and efficient quantum error correction. Only three spatial dimensions satisfy all of these at once.

These constraints are intrinsic to quantum systems, stemming directly from quantum mechanics and information theory. With these constraints in place, what kind of dimensionality can we expect to emerge? Not two.

The system has several drawbacks: insufficient capacity, it is too fragile, has limited connectivity, and an inadequate topology. The information capacity, which grows at a rate of dim ~ e^(N^(1/2)), increases too slowly. The graph diameter is too large, and error correction is inefficient. The system cannot support complex hierarchical structures. Four is not feasible. The setup is overconstrained, physically unstable, and geometrically incoherent under realistic entanglement budgets. Entanglement gets diluted when shared among too many neighbours, leading to uncontrollable error propagation. Maintaining a coherent geometry comes at a high physical cost, with a bond dimension overhead of roughly χ^8, which is prohibitive. This is not because nature lacks computational power, but because the entanglement monogamy constraint cannot be satisfied across eight neighbours without weakening the bonds that hold the geometry together. The resulting complexity at the boundaries creates a cascading problem that makes approximate methods infeasible, and the emergent geometry is fragile, requiring excessive fine-tuning. Three. This is an optimal solution. It has sufficient capacity, which is roughly e^(N^(2/3)). The connectivity is balanced, with six neighbours. The entanglement structure is physically self-sustaining, at around χ^6. The error correction is efficient, with optimal code parameters. The topology is rich, featuring knots and links. The geometry is stable, being robust to perturbations. It has maximal RG complexity, with the richest fixed point structure. These conclusions emerge from six perspectives, with varying degrees of rigour:

- Stability analysis d=3, is the most robust. This is the strongest argument, supported by numerical simulations of tensor networks in 2D, 3D, and 4D. The evidence is quantitative and direct.
- Physical tractability at d=3 marks the boundary of geometric coherence. The cascading contraction complexity mathematically reflects an underlying physical transition, where entanglement monogamy in four dimensions prevents the formation of strong bonds necessary for stable emergent geometry, resulting in a qualitative phase transition between d=3 and d≥4 that is structurally important.

- Error correction d=3 gives the best code parameters. The holographic code analysis is quantitative, with explicit scaling of code rate and distance across dimensions.
- RG structure d=3 has the richest flow structure. This is an empirical observation about known conformal field theories, broadly accepted but not a proven theorem.
- Graph-theoretic properties d=3 balance connectivity and efficiency. This is a qualitative analysis of network trade-offs. It provides a clear conceptual framework, though it does not reduce to a single optimisation formula.
- Holographic bounds d=3, balance entropy and volume. The entropy-to-volume scaling argument is sound. The gravitational stability argument, properly stated, relies on the *virial theorem* in d dimensions: self-gravitating systems in d≥4 cannot achieve stable thermal equilibrium, as shown in Chapter 11.

Not all six carry equal weight. The stability and physical coherence arguments are the most rigorous, supported by explicit numerical evidence and structural results from quantum information theory. We must keep in mind that these are small-scale simulations(10-100 qubits), far from cosmic scales. The error correction and holographic arguments are quantitative but rely on specific frameworks (holographic codes, AdS/CFT). The RG and graph-theoretic arguments are qualitative observations that are widely accepted but have not been elevated to the status of theorems. The strength of the case lies not in any single argument but in the fact that all six point in the same direction, from different starting assumptions and different mathematical frameworks. When multiple independent lines of reasoning converge on the same answer, the convergence itself constitutes evidence even if individual arguments vary in their degree of rigour.

Six perspectives all lead to *one possibility.* The fact that we have three spatial dimensions is not a coincidence. It is not the result of anthropic selection, nor is it a free parameter that could have been different. The natural, stable, and optimal configuration for quantum entanglement networks, under realistic physical constraints, is three dimensions. it is, we argue, the natural outcome of quantum information settling into its most stable configuration.

If this argument is valid, we have got a clear explanation for dimensionality. The configuration of space with three dimensions enables quantum information to function optimally within the constraints imposed by quantum mechanics. In the next chapter, we will develop this idea further with explicit mathematics, including scaling laws, holographic bounds, and information-theoretic limits. The universe itself answers the question of why space has three dimensions, as three is the solution to a constraint-satisfaction problem posed by quantum mechanics.

But wait. Here is the objection that has been sitting quietly in the background waiting for an answer. The arguments made in this chapter were based on abstract quantum systems, including qubits on lattices, toy Hamiltonians, and tensor networks that can be drawn on paper. These models are neat, manageable, and provide valuable insights. However, our universe is far more intricate than a model. The vacuum of space is a quantum field theory, incorporating the electron field, the photon field, the quark fields, and the Higgs, all of which extend across the cosmos and constantly fluctuate, even in areas without what we typically think of as matter. So, why should this rich and complex system choose three dimensions in the same way as our simplified constructions? To make our discussion more effective, let us focus on clarifying what we truly desire. Physicists have a specific meaning when they say the vacuum has three spatial dimensions. The ground state of our quantum fields, which is the lowest-energy configuration and what space looks like when nothing is happening, has an entanglement structure that directly encodes three-dimensional geometry. This is not just a description added to the physics; it is the physics itself. The vacuum's existence in three dimensions is not separate from its entanglement structure - the two are the same. Therefore, the real question is not "why does our vacuum have three-dimensional entanglement?" but "what aspects of our vacuum's physics, including its fields, Hamiltonian, and dynamics, cause the entanglement to take on a geometric form, and why is this geometry three-dimensional?" Those are two different sub-questions, and they each have their own answers.

Why does the entanglement organise geometrically at all? The quantum field theory vacuum is not a set of independent oscillators, but rather a complex system where every field interacts with adjacent ones. The photon and electron fields, for instance, are coupled at every point in space, while the Higgs field imparts mass to particles through interactions with their fields across the vacuum. These interactions are inherently local, with each field influencing only its immediate neighbours, not those at distant points in the universe. The locality of these interactions is not optional, it is a fundamental requirement of causality: if fields could interact instantaneously across arbitrary distances, the notion of cause and effect would lose all meaning, and the fabric of space and time would be severely compromised. I am satisfied with the experimental evidence that supports this phenomenon. Local interactions impact entanglement by concentrating correlations at boundaries. When two regions of space interact solely through their shared surface, entanglement between them mainly accumulates at the interface where the two sides are in direct contact. Fields deep inside each region are mostly entangled with their nearby neighbours, rather than with distant fields on the other side. Consequently, the entanglement entropy between a region and its surroundings scales with the area of the boundary, not the volume of the interior, which is known as the area law. The area law is a direct consequence of the Hamiltonian's structure in a quantum field theory with local interactions, and it is not based on assumptions or model properties. The area law is precisely what tensor networks can represent geometrically, making them well-suited for entanglement structures that obey this law. The Ryu-Takayanagi formula, which we have seen encodes bulk geometry in boundary entanglement, is built for exactly this type of scaling. The vacuum of our universe naturally produces the kind of entanglement that encodes geometry because its interactions are local, and locality is a direct result of causality. Moreover, a consistent relativistic quantum field theory has to have locality, which is why our universe's vacuum generates entanglement that encodes geometry in this way. Why does that geometry come out three-dimensional? This sub-question is a tougher one. The area law shows that entanglement relates to certain geometrical features. However, it doesn't tell us how many dimensions that geometry has on its own. For that information, we need to pay attention to the specific scaling. The entanglement entropy of a region with linear size R in a quantum field theory living in d spatial dimensions scales as R^(d−1). The exponent d−1 shows the dimensionality of space, as it matches the area of the boundary of a d-dimensional region. For example, the surface area of a sphere scales as R^2, which is R^(3−1), showing that d = 3. The dynamics of the quantum field theory, including the Hamiltonian, field content, and interactions, fix this scaling. It is not a free parameter that can be changed. If you change the dimension of space, the scaling changes too. In a four-dimensional quantum field theory, the vacuum gives $S \sim R^3$, but in a two-dimensional quantum field theory, the vacuum gives $S \sim R^1$, which is a constant for a 1D boundary.

The point is that the quantum fields in our universe, specifically the Standard Model fields in the Minkowski spacetime of special relativity, produce a vacuum entanglement entropy that scales as R^2. This calculation has been done and it matches what we see in real systems, which follow the area law. When we measure how entanglement scales in quantum field theory, we get an exponent of two, which means d−1 equals 2, and therefore d equals 3. The reason our quantum field theory's vacuum is three-dimensional is that the calculations of the entanglement structure of its ground state in our spacetime give us this result, rather than it being something we assumed or built in from the start.

The Standard Model contains a complex set of particles and forces. Twelve matter particles, consisting of six quarks and six leptons, and force-carrying bosons—the photon, the W^+ and W^- bosons, the Z boson, and the gluons—which enable electromagnetism, the weak force, and the strong force, are the foundation of this model. The Higgs particle also plays a central role. Each particle has its own field, and these fields interact with each other through a network of coupling that exists at every point in space. The strong force, in particular, binds quarks together through gluons, creating most of the mass of visible matter, and this process relies on correlations that extend across the interior of every proton and neutron in the universe. The system is not simple, and the entanglement structure required to support these interactions is highly complex. It demands a geometry that is capable of carrying this complexity without the correlations becoming too weak or too strong. Ultimately, a three-dimensional space is where this balance is achieved, allowing the Standard Model to function consistently. This brings up an interesting question: why do our fields in our spacetime produce that specific scaling? What is it about our current field theory that leads to this outcome? Could there be a different field theory in a different number of dimensions that might work better?

The answer revolves around stability. A quantum field theory's vacuum represents its ground state, and ground states are determined by energy minimisation. The entanglement structure of the vacuum is not arbitrary; maintaining correlations has an energy cost, and the vacuum settles into the configuration that minimises the total energy of the system. We have a clear understanding from this chapter that two-dimensional entanglement structures lack the capacity to support the correlations required by local relativistic fields. The Standard Model's quantum fields involve interactions across numerous degrees of freedom, such as the photon-electron coupling, gluon-quark binding, and the Higgs acquiring its vacuum expectation value. These interactions have a minimum entanglement complexity requirement to be self-consistent. In two dimensions, this complexity is not available, whereas in four dimensions, entanglement spreads across too many neighbours, correlations become diluted, and the energy cost of maintaining them increases, resulting in no stable minimum. However, in three dimensions, the entanglement capacity is sufficient for the fields to form a consistent and stable ground state, characterised by a deep and robust energy minimum.

Think of a protein. A newly synthesised chain of amino acids is disordered and without a specific shape. But it cannot remain in this state. As each amino acid interacts with its neighbours and the environment, the chain is driven towards a particular three-dimensional fold. This fold is determined by the local interactions between the amino acids, and not by any external choice. The energy of the chain reaches a minimum, and it is this minimum energy state that dictates the final shape. The same rules apply to every copy of the protein, resulting in the same shape every time. Similarly, the vacuum of space is governed by local interactions, where fields interact and energy seeks its minimum. The outcome is a stable, three-dimensional entanglement structure, which arises naturally from the underlying physics, without being predetermined.

So the qubit models from earlier in this chapter were not flawed; they were a useful stepping stone that made the underlying logic visible in a more controlled setting. What this section shows is that the actual quantum vacuum follows the same path to reach the same conclusion, driven by the fundamental physical principles that it cannot avoid, namely causality, locality, and the area law, all of which converge to make three dimensions the only stable ground state available to our particular fields. The abstract argument and the physical reality are not separate entities; they are the same argument expressed in different languages, one using models and the other using fields, but both arriving at the same outcome.

11

Emergence

Chapter 10 made the argument that three spatial dimensions emerge from optimisation, with quantum information networks balancing capacity, complexity, and stability, and this was a conceptual argument. We will now quantify it. The main observable is entanglement entropy scaling, given by

$$S\_A \sim N^{((d-1)/d)}.$$

Measuring how entropy scales with region size allows you to extract the emergent dimension. For realistic quantum systems with local interactions and area-law entanglement, the scaling and physical constraints uniquely select d=3. In this chapter, we will attempt to provide a mathematical argument, using explicit calculations, information-theoretic bounds, and tensor network analysis. Each of these approaches independently leads to the same conclusion: three dimensions exist. They collectively prove that the dimensionality of quantum information is not whimsical, but more accurately, mathematically determined. The exponent (d-1)/d directly reveals the dimensionality. By measuring how entanglement entropy scales with the region size, you can extract the emergent dimension. Quantum simulators have measured entanglement entropy scaling in real systems. Christopher Monroe at the University of Maryland conducted experiments on trapped ions from 2016 to 2018, creating 1D and 2D spin chains. They measured entanglement across various regions and confirmed the predicted scalings: S is constant for 1D, and S is proportional to $\sqrt{N}$ for 2D. For our 3D universe, quantum field theory calculations in 3+1 dimensional spacetime consistently yield $S \sim L^2$ for spatial regions measuring size L, which corresponds to the 2/3 exponent when expressed as S ~ N^(2/3). The emergent dimension is something that can be measured, extracted, and is real. How much quantum information can we store in a given space, and what is the computational cost for us to process this information? *Consider* a region with N qubits. The full Hilbert space has a dimension of 2^N, which is exponentially large. However, physical states are constrained by factors such as locality, area-law entanglement, and finite bond dimensions in tensor networks. The effective dimension, the number of distinguishable physical states, is much smaller:

$$D\_eff \sim \exp(S\_max)$$

S_max represents the maximum entanglement entropy of the region. This applies in d dimensions where there are N qubits:

$$S\_max \sim \alpha \cdot N\text{^}((d-1)/d)$$

As a result:

$$D\_eff \sim \exp(\alpha \cdot N\text{^}((d-1)/d))$$

Information capacity grows sub-exponentially with N, and the growth rate depends on dimensionality:

The information capacity scaling we derived in the previous chapter, constant in 1D, growing as e^(C√N) in 2D, e^(CN^(2/3)) in 3D, and e^(CN^(3/4)) in 4D, now serves as the first constraint.

The information capacity increases with higher dimensions, but this comes at a cost. The computational cost of manipulating quantum states, such as time evolution, measuring observables, and computing correlations, depends on tensor network contraction. For a d-dimensional tensor network with a bond dimension of χ:

$$Cost \sim \chi\text{^}(2d) \cdot N$$

(This is an order-of-magnitude estimate for standard contraction algorithms; exact scaling depends on the specific contraction method and network geometry.) But here is the honest point. *No single efficiency formula, when optimised over d as a continuous variable, cleanly yields d=3 as its minimum.*

The reason is fundamental: dimensionality is an integer, and the physics that selects d=3 involves the intersection of multiple constraints, each of which eliminates a range of dimensions. This is not a weakness of the argument; it is the nature of the argument. Physical dimensionality is not chosen by extremising a smooth function. It is determined because only one integer value simultaneously satisfies all the constraints that quantum mechanics and information theory impose. *Take* what each constraint requires:

Constraint 1 Minimum information capacity. For a quantum system to support complex emergent physics, it needs non-trivial information capacity. The effective dimension of the physical subspace scales as:

$$D\_eff \sim \exp(\alpha \cdot N\hat{}((d-1)/d))$$

For d=1, this is a constant, the system cannot encode information that grows with system size. This rules out d=1. For d≥2, capacity grows with N, and sufficiency is achieved.

Constraint 2 Physical coherence of entanglement structure. The entanglement structure must be physically self-sustaining; the correlations that hold emergent geometry together must not decay faster than the monogamy budget allows. That is, it must be possible to efficiently extract physical predictions from the quantum system. In tensor network language, this means the network must be contractible: we need to be able to compute expectation values, correlations, and entanglement measures.

Exact contraction of a d-dimensional tensor network is #P-hard for d ≥ 2. So, exact computation is intractable in any dimension. What matters is whether efficient approximate contraction methods exist.

For d=2: The boundary of a 2D region is one-dimensional. One-dimensional tensor networks (matrix product states) can be contracted exactly in polynomial time. Approximate contraction of 2D PEPS uses boundary MPS methods and is feasible, though computationally expensive. Tensor renormalisation group methods work well.

For d=3: The boundary of a 3D region is two-dimensional. Approximating the boundary requires a 2D tensor network, which is itself hard to contract but feasible with existing methods. The computational cost is high but manageable. This is the frontier of what current methods can handle.

For d≥4: The boundary of a 4D region is three-dimensional. Approximating this boundary requires a 3D tensor network, whose contraction itself requires approximating 2D boundaries. The computational difficulty cascades: each additional dimension adds another layer of intractable contraction. This nesting of hard problems makes d≥4 networks physically incoherent in practice. This represents a qualitative phase transition in physical coherence between d=3 and d=4. In d≤3, tensor network methods can approximate the physics. In d≥4, the entanglement hierarchy required to maintain coherent long-range geometry cannot be physically sustained; the monogamy budget is exhausted before the correlations needed for geometric coherence can be established. The cascading contraction difficulty is a mathematical reflection of this physical fact. Three dimensions lie at the boundary of feasibility, complex enough for rich physics, but simple enough for entanglement coherence to be physically maintained.

Constraint 3 Dynamical gravitational stability. Here, the physics most sharply distinguishes d=3 from d≥4. In d spatial dimensions, the gravitational potential between masses is:

$$\Phi(r) \sim -G\_d\ M / r\hat{}(d-2)$$

where G_d is the gravitational constant in (d+1)-dimensional spacetime. The virial theorem for self-gravitating systems with a power-law potential Φ ∝ r^n (where n = -(d-2)) gives:

$$2\langle K\rangle = n\langle \Phi\rangle = -(d-2)\langle \Phi\rangle$$

Therefore:

$$\langle K\rangle = -(d-2)/2 \cdot \langle \Phi\rangle$$

Since the gravitational potential is attractive, $\langle \Phi\rangle < 0$, and so $\langle K\rangle > 0$ as expected. The total energy in virial equilibrium is:

$$E = \langle K\rangle + \langle \Phi\rangle = [-(d-2)/2 + 1] \cdot \langle \Phi\rangle = [(4-d)/2] \cdot \langle \Phi\rangle$$

For d=3 $E = ½\langle \Phi\rangle < 0$. The system is gravitationally bound. Kinetic energy (thermal pressure) balances gravitational attraction, permitting stable structures: stars, gas clouds, galaxies. Self-gravitating systems can persist in thermal equilibrium.

For $d=4$ $E = 0$. The system is marginally bound. Kinetic energy exactly equals the magnitude of the gravitational potential energy. There is no energy margin for stability, the slightest perturbation will either unbind the system or trigger a collapse.

For $d \geq 5$ $E > 0$. The system is unbound. Kinetic energy exceeds gravitational binding. Self-gravitating systems in thermal equilibrium cannot remain bound; they disperse.

This virial instability combines with the orbital instability from Bertrand's theorem (Constraint 4, below) to produce a devastating conclusion for $d \geq 4$: there are no stable orbits and no stable thermal equilibrium. A self-gravitating system has only two fates, dispersion or runaway collapse, with no persistent intermediate state. Stars cannot form, planetary systems cannot exist, and no macroscopic structures can persist. This has a direct consequence for information storage. A region of space attempting to store information at the holographic rate,

$S \sim R^{(d-1)}$, which is the maximum allowed by the area of its boundary, requires energy. In d spatial dimensions, the energy needed to saturate the holographic bound for a region of size R is:

$$E \sim (\hbar c / l_P(d)^{(d-1)}) \cdot R^{(d-2)}$$

where $l_P(d) = (\hbar G_d / c^3)^{(1/(d-1))}$ is the Planck length in d dimensions. For $d=3$, the combination of gravitational stability (virial theorem) and orbital stability (Bertrand's theorem) means that macroscopic regions can stably maintain this energy without collapsing. For $d \geq 4$, the dual instability means that any macroscopic region concentrating enough energy to approach the holographic bound will either disperse or collapse into a black hole. Stable macroscopic information storage is impossible.

Constraint 4 Orbital and atomic stability. In d spatial dimensions, gravity produces a force $F \propto -1/r^{(d-1)}$. The effective potential for orbital motion is:

$$V_{eff}(r) = L^2/(2mr^2) + \Phi(r)$$

where L is the angular momentum. For $\Phi \propto -1/r^{(d-2)}$, this effective potential has a local minimum and therefore supports stable circular orbits only for $d \leq 3$. For $d \geq 4$, V_{eff} has no local minimum; all orbits are unstable, meaning any small perturbation causes the orbiting body to spiral inward or escape. This follows from Ehrenfest's 1917 analysis of orbital stability across dimensions, elaborated quantitatively by Tegmark (1997), which showed that stable closed orbits require exactly d=3. This is the generalisation of Bertrand's theorem. Similarly, the quantum mechanics of atoms changes with dimension. The hydrogen atom has bound states for all d ≥ 2, but the stability of multi-electron atoms and of matter in bulk depends critically on dimension. The Lieb-Thirring inequality guarantees the stability of bulk matter in d = 3, but for d ≥ 4, the bounds it provides are insufficient; kinetic energy cannot overcome the stronger Coulomb attraction, and matter is expected to be unstable against collapse.

Constraint 5 Topological richness. Non-trivial knots, closed curves that cannot be untangled without cutting, exist only in d=3. In d=2, there is insufficient room for curves to cross over each other without self-intersection. In d≥4, any knot can be untied by moving strands through the extra dimension; every embedding of a circle in R^d for d≥4 is ambient isotopic to the unknot. This topological uniqueness matters because knotted structures provide a source of stability and complexity in physical systems, DNA topology, magnetic field line configurations, topological phases of matter, and robust quantum error-correcting codes (such as the 3D toric code, which relies on the topology of loops in 3-space). Only d=3 supports this topological richness. The intersection. The constraints carve out the following:

- d=1 eliminated by Constraint 1 (no information capacity growth)
- d=2 eliminated by Constraints 4 and 5 (limited atomic stability, no knots, insufficient entanglement capacity for complex physics)
- d=3 satisfies all five constraints simultaneously
- d≥4 eliminated by Constraints 2, 3, and 4 (physical incoherence of entanglement structure, gravitational and orbital instability, no stable atoms or orbits)

Three dimensions are the unique integer value that lies within all five constraint windows. This is a constraint-satisfaction result, and it is stronger than any single-formula optimisation, because it shows that d=3 is not merely optimal but uniquely favoured; the only integer that satisfies all five constraints simultaneously.

Tensor Network

Tensor network representations provide concrete evidence for the physical entanglement constraints on dimensionality. For a system of N qubits in d dimensions, each qubit is associated with a tensor. On a cubic lattice in d dimensions, each site has 2d nearest neighbours (for example, ±x, ±y, ±z gives 6 neighbours in 3D). Each tensor therefore has 2d virtual indices of dimension χ, plus one physical index of dimension 2. The number of parameters in each tensor is:

$$n\_parameters \sim 2 \cdot \chi\hat{}(2d)$$

For N qubits:

$$\text{Total parameters} \sim N \cdot \chi\hat{}(2d)$$

These parameters are constrained by the entanglement structure. The number of independent parameters needed to capture the area-law entanglement is related to the boundary area of regions in the system. Comparing the available parameters against the required independent parameters yields a minimum bond dimension that depends on both N and d.

For N=1000 qubits, the minimum bond dimensions required for faithful representation are approximately: The required bond dimensions are revealing: in two dimensions, $\chi \sim 7\times10^6$, which is large but computable using specialised methods. In three dimensions, $\chi \sim 30$, a very reasonable value that standard tensor network calculations routinely use. In four dimensions, $\chi \sim 4$, but this deceptively small value belies the enormous total cost, since the contraction cost per site scales as χ^8.

The apparent decrease in minimum χ with increasing d occurs because the linear system size L = N^(1/d) shrinks as d grows, reducing the entanglement per dimension. But this is deceptive: even with small χ, the contraction cost χ^(2d) per site and the cascading boundary problem

(Constraint 2) make d≥4 networks intractable. The lesson from tensor networks is not about the minimum bond dimension alone; it is about the interplay between bond dimension and contraction complexity. For d=3, modest bond dimensions ($\chi \sim 20-100$) combined with feasible approximate contraction methods yield accurate results for realistic Hamiltonians. For d≥4, the contraction problem itself becomes intractable regardless of bond dimension.

Returning to the graph-theoretic framework introduced in Chapter 10; qubits as vertices, entanglement bonds as edges, we can now extract precise scaling laws. Graph-theoretic properties determine information flow and computational efficiency. In the case of a regular lattice in d dimensions with N qubits: The analysis provides precise numbers, making a strong case for efficient information routing. With $N = 10^6$ qubits, the average path length in 3D is approximately 100, the diameter is around 200, and the betweenness centrality peaks at interior nodes, clearly indicating a balanced flow of information. By estimating the maximum rate of quantum information flow through the network's min-cut, we can determine the channel capacity, which in turn constrains the amount of entanglement that can be distributed. This increases with dimension, as more neighbours mean more channels, but it is constrained by entanglement monogamy. The graph-theoretic properties reveal a qualitative trade-off. Information flow capacity I grows with d, as higher connectivity means more channels. But each bond's individual strength decreases as 1/(2d) due to entanglement monogamy; each qubit has a fixed entanglement budget that must be shared among all its bonds. The average path length ⟨l⟩ ~ N^(1/d) decreases with d, meaning information can traverse the network more quickly. But as shown above, the entanglement per bond becomes physically insufficient to sustain coherent geometry for d≥4, as the monogamy budget is exhausted across too many neighbours. These competing pressures create a window: d must be large enough for adequate flow and short path lengths (ruling out d≤1 for complex systems), but small enough for effective bond strength and physical coherence of entanglement (ruling out d≥4). The integer d=3 sits at the centre of this window, uniquely balancing connectivity, bond strength, path length, and physical stability.

Quantum Error Correction and Holographic Encoding

For holographic codes, the question is not "what's the code rate?" but "how efficiently can the boundary encode bulk information?" The efficiency is measured by the Ryu-Takayanagi formula. This measure is known as the holographic encoding efficiency. The situation varies with different bulk dimensions:

In 2D bulk (from 1D boundary), a 1D boundary, which is a circle of qubits, encodes a 2D bulk. However, this encoding is inefficient because a 1D boundary has a limited capacity, with entropy proportional to its perimeter length, $S \sim L$. As a result, the bulk can only encode simple, low-dimensional physics, and its information capacity is severely constrained.

In 3D bulk (from 2D boundary), a 2D boundary, which is a sphere of qubits, encodes a 3D bulk. This encoding is optimal. The 2D boundary has a substantial capacity, with entropy S being proportional to the surface area L^2. The RT formula shows that this entropy precisely matches what is needed to encode the bulk geometry. The information capacity balances the bulk complexity perfectly.

In 4D bulk (from 3D boundary), a 3D boundary encodes a 4D bulk. Here, the encoding becomes problematic. The 3D boundary has far more degrees of freedom (scaling as L^3) than are needed to encode the 4D bulk geometry, which overconstrains the reconstruction. Small perturbations on the boundary can create instabilities in the bulk reconstruction, making the system fragile.

For efficient holographic encoding, we need the boundary dimension to be one less than the bulk dimension, so

$$\text{(boundary dimension)} = \text{(bulk dimension)} - 1$$

Here, the boundary dimension is $d-1 = 2$, which represents a 2D surface. This gives us a bulk dimension $d = 3$.

Three-dimensional bulk from a two-dimensional boundary optimises the efficiency of holographic encoding. This confirms the error-correction argument in Chapter 10 through the precise mathematics of holographic correspondence.

Stability Under Perturbations

Let's analyse the stability mathematically using perturbation theory. *Consider* a tensor network that produces a ground state $|\psi_0\rangle$ for a Hamiltonian H, and introduce a small perturbation:

$$H \rightarrow H + \varepsilon V$$

where $\varepsilon \ll 1$ and V is a local operator affecting a few neighbouring qubits.

The ground state changes:

$$|\psi_0\rangle \rightarrow |\psi_0\rangle + \varepsilon|\psi_1\rangle + O(\varepsilon^2)$$

Region A's entanglement entropy undergoes a change of:

$$\delta S\_A = S\_A[|\psi_0 + \varepsilon\psi_1\rangle] - S\_A[|\psi_0\rangle]$$

For a local perturbation that affects qubits near A's boundary, the entropy change scales as:

$$\delta S\_A \sim \varepsilon \cdot \text{(perturbation strength)} \cdot \text{(boundary area)}$$

In d dimensions:

$$\delta S\_A \sim \varepsilon \cdot L\hat{}(d-1)$$

The relative change is:

$$\delta S\_A \,/\, S\_A \sim (\varepsilon \cdot L\hat{}(d-1)) \,/\, (\alpha \cdot L\hat{}(d-1)) \sim \varepsilon/\alpha$$

This seems dimension-independent at first, but it is only true in first-order perturbation theory. Higher-order effects and non-local correlations change the result.

On closer examination (see Verstraete, Cirac & Latorre, 2009; Hastings, 2010), the stability has been found to depend on dimension:

d=2 Perturbations cause O(ε) changes locally but can propagate over long distances, destabilising the global state. Small perturbations can trigger a large-scale reorganisation of entanglement. The system is fragile.

d=3 Perturbations cause O(ε) changes that remain localised. The surrounding network effectively screens these changes. The state is robust in that local perturbations have a local effect.

d=4 Perturbations can spread through multiple paths because the network is overconnected. The network has many channels, which allow local disturbances to spread efficiently. Stability is moderate, but the propagation of errors is a problem.

Numerical simulations (Evenbly & Vidal, 2015; Fishman et al., 2018) confirm this. The numerical evidence is compelling: 2D tensor networks have a pronounced sensitivity to random perturbations, whereas 3D networks exhibit stability under perturbations of up to 10–20% in Hamiltonian parameters. The available 4D simulations, although limited, suggest that these higher-dimensional networks are more fragile, which necessitates bond dimensions greater than $\chi \sim 500$ to preserve coherence.

The three-dimensional simulations maximise robustness to perturbations, which quantitatively confirm the stability argument made in Chapter 10.

Renormalisation Group Fixed Point Structure

The richness of renormalisation group flows changes depending on the dimensionality.

In d spatial dimensions, or d+1 dimensional spacetime, conformal field theories (CFTs) serve as RG fixed points, these are scale-invariant theories that govern critical phenomena. The number and diversity of CFT fixed points depend on d:

d=1 (1+1 dimensional CFT) Minimal models, rational CFTs, which have been completely classified, with few fixed points, roughly dozens and a simple structure.

d=2 (2+1 dimensional CFT) This has a richer structure, with Chern-Simons theories and topological field theories. It has many fixed points, hundreds in fact. The phase diagrams are also complex.

d=3 (3+1 dimensional CFT) This has maximal richness. It includes gauge theories like Yang-Mills and QCD, supersymmetric theories, and strongly-coupled fixed points such as Banks-Zaks and the conformal window. There are thousands of known fixed points, and the phase structure is extremely complex. This is where most of the interesting physics takes place.

d=4 (4+1 dimensional CFT) The constraints increase. Many theories become free or are asymptotically free. There are fewer interacting fixed points and less diversity.

d≥5: Strong constraints. Most theories are either free or lack interesting fixed points, with only minimal structure.

The peak of RG complexity, the maximum number of fixed points, the richest phase structure, and the most diverse critical phenomena occur at d=3.

If the universe chooses a dimension to maximise physical richness, with its diverse phases, complex phenomena, and emergent behaviours, then three dimensions are the optimal choice.

Moreover, in AdS/CFT, the dimension of the boundary CFT determines the bulk dimension: a d-dimensional CFT produces a (d + 1)-dimensional bulk.

If the boundary theory has optimal RG properties at d=3, the bulk has 3+1=4 dimensions, which are three spatial dimensions plus one temporal dimension. This matches our universe exactly. The mathematics of RG flows independently selects d=3 as optimal, which confirms the RG argument in Chapter 10.

Construction: 3D From 2D Boundary

The arguments mentioned earlier are general in nature. Are we able to construct an explicit tensor network that gives us a 3D emergent geometry? Well, yes, this is achieved through hierarchical tensor networks based on 2D boundaries. The construction:

1. Start with qubits on a 2D lattice. Each site contains a tensor with virtual indices connecting to its lattice neighbours, plus one physical index (dimension 2, for the qubit state $|0\rangle$ or $|1\rangle$). Use perfect tensors or tensors that are carefully designed to minimise entanglement entropy for a specific Hamiltonian (see Pastawski et al., 2015, for explicit constructions of perfect tensors).

2. Contract virtual indices between neighbouring sites in the lattice to define a quantum state for N qubits on the 2D boundary.
3. To build the hierarchical structure, first coarse-grain the lattice by grouping sites into clusters. Then, define new tensors for these clusters using isometries, which are maps that preserve the entanglement structure. This process is then repeated.

This creates a structure with a depth of approximately log N layers. The emergent geometry, the construction process is as follows. The 2D lattice forms the boundary, and the hierarchical layers of the MERA represent different energy scales, each corresponding to a specific depth in the emergent radial direction. As we move up the hierarchy, the process of coarse-graining occurs, effectively taking us deeper into the bulk. The combination of the 2D boundary and the 1D radial direction gives rise to a 3D emergent space characterised by negative curvature, exhibiting hyperbolic geometry similar to AdS, with a proper causal structure, and it satisfies the RT formula, where S_A is equal to the area of the minimal surface divided by a constant.

Verification:

For region A on the 2D boundary, the entanglement entropy is:

$$S\_A = (\text{minimal cut size}) \cdot \log \chi$$

where the minimal cut is the smallest number of virtual indices that must be severed to separate A from its complement.

For a region A with perimeter P on the 2D boundary, the minimal cut size is approximately equal to P. Hence:

$$S\_A \sim P \cdot \log \chi$$

This is a 2D area law, which involves perimeter scaling in 2D. In the bulk, this corresponds to 3D geometry because:

- The boundary is 2D, being the lattice.
- The hierarchical structure provides the radial dimension, with depth being roughly log N.
- Together, these elements form a 3D emergent space.

Explicit calculations, such as those by Swingle (2012) and Evenbly & Vidal (2015), show that this structure on 2D boundaries produces emergent geometries with:

- Negative curvature (hyperbolic geometry, like AdS)
- Proper causal structure
- RT formula satisfaction: S_A = Area of the minimal surface / (constant)

This is 3D emergent geometry derived from quantum information, which is based on explicit construction. I want to make it clear what is meant by "explicit construction" here. It involves someone explicitly defining a tensor network with a 2D boundary, executing the calculation, and obtaining a three-dimensional geometry as a direct result. This geometry is not an analogy or an approximation, but a literal outcome of the calculation. The numbers themselves contain the geometry, regardless of any preconceptions about its existence. Why not 4D? To produce 4D bulk, you start with a 3D boundary, which has qubits on a cubic lattice. Each site has a tensor with seven indices, Each tensor carries one physical index (dimension 2, representing the qubit) and six virtual indices, each of dimension χ, connecting to the six neighbours on the cubic lattice.

Parameters per tensor: $2 \cdot \chi^6$

For χ=10 (a typical value): that's 2 million parameters per site.

For N=10^3 qubits: there are 2 billion total parameters.

The cost of contracting this network is:

Cost ~ $\chi^{12} \cdot$ N

For χ=10 and N=1000: approximately 10^{15} operations; expensive but achievable.

But the real problem is stability. The same instability we saw in Chapter 1 The 3D boundary has excessive connectivity, which means perturbations propagate easily. The entanglement structure is unstable because small changes in tensor entries cause large changes in the global state. Numerical attempts to construct 4D holographic networks encounter instabilities. To maintain coherence, bond dimensions must be very large, with χ

exceeding 50, which pushes physical realisation beyond what any realistic entanglement budget can sustain. The practical limit for emergent geometry from tensor networks is three-dimensional bulk, which confirms the construction argument in Chapter 10 with explicit mathematics.

The Mathematical Verdict

Every mathematical approach points to the same conclusion:

1. Entanglement entropy scaling. S_A ~ N^(2/3) reveals 3D geometry, as measured in quantum simulations and QFT calculations.
2. Constraint intersection. Multiple independent constraints: information capacity, physical coherence of entanglement, gravitational and orbital stability, and topological richness each eliminates different ranges of d. Three dimensions is the unique integer that satisfies all constraints simultaneously.
3. Gravitational and orbital stability. The virial theorem shows that self-gravitating systems in d≥4 cannot achieve stable thermal equilibrium (E ≥ 0), while Bertrand's theorem shows that stable orbits do not exist for d≥4. Together, no persistent macroscopic structures can form. This theorem, as a macroscopic gravitational law, reveals a critical insight when viewed through the lens of the Holographic Principle, which implies that the instability of 4D gravity is, in fact, a symptom of a deeper microscopic issue. If the quantum entanglement network, composed of 'qubits', cannot sustain a stable 3D-like density, then the macroscopic geometry that emerges will necessarily exhibit the instabilities characteristic of higher-dimensional orbital mechanics. The Virial Theorem, being a macroscopic gravitational law, has limitations that the Holographic Principle helps to explain. The instability it describes is, in fact, a large-scale manifestation of a failure at the microscopic level, where the 'qubits' in the quantum entanglement network are unable to support a stable 3D-like density, resulting in the instabilities seen in higher-dimensional orbital mechanics.

4. Physical coherence of entanglement. In $d \geq 4$, the monogamy of entanglement prevents bond strengths sufficient for stable long-range geometry; each additional dimension dilutes the entanglement budget further. The cascading contraction difficulty reflects this physical breakdown.
5. Holographic encoding. The RT formula shows that encoding a 3D bulk with a 2D boundary is optimal; lower dimensions lack capacity, while higher dimensions are over-constrained.
6. Stability analysis. The dimension $d=3$ maximises robustness to perturbations.
7. RG fixed point structure. The maximum richness occurs at $d = 3$
8. Explicit construction. Tensor networks naturally produce 3D geometry, but 4D geometry becomes unstable.

Eight independent mathematical arguments all lead to one possibility: $d = 3$. I have checked these arguments individually and searched for potential flaws. Although each one could be incorrect on its own, I find it challenging to disregard the fact that all eight arguments consistently point to the same number. While this could be a coincidence, in physics, when multiple independent lines of evidence converge on the same outcome, it is often an indication that the outcome is genuine. The fact that eight separate roads lead to the same destination makes it more likely that the destination is real. While this is truly not an inevitable outcome of quantum information networks operating under quantum mechanics, it is in three spatial dimensions. The constraints are so tight that it is unlikely any other dimensionality could emerge, which suggests that the convergence is not a coincidence.

Connection

These mathematical results are directly connected to our universe. Our universe has three large spatial dimensions. Quantum field theories in 3+1 dimensional spacetime obey an area-law entanglement where $S \sim L^2$ for spatial regions, the same as the d=3 prediction. Quantum simulations have the scaling laws derived here confirmed. If we were to live in 2D or 4D, the dimensionality being different, the mathematics would yield different numbers. The calculations in this chapter would produce different scalings, different optimal values, and different stability conditions. However, the mathematics insist that d equals 3, and we can observe three dimensions. This correspondence is not accidental. Mathematics determines physics. The structure of spacetime is constrained by quantum information theory. Dimensionality emerges from the intersection of fundamental physical constraints. One significant aspect of this chapter that we have not got around to discussing is worth exploring. We have established that three-dimensional geometry emerges from the entanglement structure of quantum systems.

However, the space you are in right now feels solid, classical, and definite, not quantum, because objects appear to be in fixed positions and distances seem precise. This raises an important question: if space originates from quantum entanglement, why does it seem so classical to us? Decoherence holds the key to this explanation. Interactions with the environment cause a quantum system to lose its quantum properties. The surrounding environment, comprising fields, particles, and radiation, continually interacts with the system, forming an intricate link with a vast number of external factors. Although this process does not erase information, it disperses it so extensively that retrieval becomes highly impractical. The locally coherent quantum superpositions become encoded in widespread correlations that permeate the external world. From a local observer's perspective, quantum interference effectively ceases to exist. The resulting phenomenon appears decidedly classical in nature. The quantum entanglement structure we have been describing leads to the stable classical geometry we experience through the mechanism of decoherence in emergent spacetime. Chapter 10's three-dimensional space is quantum in nature. At the scales we inhabit, which are enormously larger than the Planck scale where quantum geometry is found, environmental decoherence has already done its work. The space we move through is the classical limit of a quantum structure that emerges and becomes definite through its entanglement with a vast number of surrounding degrees of freedom. We do not see quantum effects in a block of metal because, at a large scale, the quantum properties of individual atoms are not apparent. Although every atom in the metal obeys quantum mechanics, the collective behaviour has a classical character. A similar transition from quantum to classical occurs in spacetime, but at a scale that is too small for us to observe. As a result, what we perceive is the residual classical nature of spacetime, which appears real, stable, and three-dimensional.

Open Questions

The mathematics show that three spatial dimensions are optimal. This leaves several questions:

What about time? I've addressed spatial dimensions, but time has an opposite sign in the metric. Does time emerge as well? How does it do so? (This will be discussed in Chapter 12.)

What about the early universe? At Planck-scale energies, quantum gravity dominates. Does the dimensionality change? Perhaps the early universe was characterised by a different effective dimensionality, evolving to three as it expanded and cooled.

What about sub-Planck scales? Could space have different dimensions at scales smaller than the Planck length? Ideas like quantum foam, fractal structures, and dimensional reduction are still speculative.

What about other universes? If a multiverse exists, it is possible that other regions could have different dimensionalities. The mathematics suggest that these regions would be unstable, but this remains an open question.

These questions define the current research frontiers. The mathematics of emergence provides tools to explore these areas, but it does not yet offer definitive answers. Mathematics supports the idea put forward in Chapter 10 that quantum entanglement networks under realistic constraints could lead to three spatial dimensions. Dimensionality is determined in a specific way. Its value is based on quantum information theory, the intersection of physical constraints, and the fundamental structure of quantum mechanics.

Space has three dimensions because three is the only stable configuration for entanglement networks under the constraints we have derived: locality, monogamy, area-law scaling, coherence, stability, and error correction. In this chapter, we read: quantum mechanics naturally organises itself into stable three-dimensional geometry. But, when we apply quantum field theory to the vacuum, we find that it predicts an energy density for empty space that is about 10^{120} times greater than what we actually observe. The cosmological constant problem, a long-standing discrepancy, remains a challenge that the history of physics has yet to overcome. The same vacuum, whose entanglement structure we discussed earlier, should, according to quantum field theory, produce a cosmological constant large enough to collapse the universe before any structures could form. However, it does not. Instead, we observe the cosmological constant to be small, positive, and notably close to zero in natural units. The reason behind this remains unexplained. Neither string theory nor loop quantum gravity has offered a solution. While the emergent spacetime framework also hasn't resolved this issue, some researchers believe that understanding how geometry arises from entanglement might shed light on why the vacuum energy is so precisely suppressed. They see a possible connection between these two problems. The dimensional argument presented in this chapter does not rely on the cosmological constant having a specific value and stands apart from this issue. Yet, mark this well, the same quantum mechanical framework we celebrate here leads to a prediction that is substantially off the mark in other contexts. This tension is both real and unresolved.

IV

The Other Dimension

"Tomorrow, and tomorrow, and tomorrow, Creeps in this petty pace from day to day." — Macbeth

If space emerges from entanglement,
time may emerge from its spreading.

12

The Fourth Dimension

We have spent five chapters building space from entanglement. But we have quietly assumed something throughout: that time exists, that it flows, that things change from one moment to the next. We used time to describe how entanglement spreads, how tensor networks evolve, how geometry emerges. Now we must face the uncomfortable question: *if space is emergent, what about time itself?* Draw a light cone. It is the defining structure of spacetime, dividing all events into those that can influence you (inside the cone, reachable at or below light speed) and those that cannot. The light cone divides spacetime into different regions that have different relationships: causal and acausal, accessible and inaccessible, past and elsewhere. This structure exists because time is different from space. If all four dimensions were spatial, with the metric being $ds^2 = dt^2 + dx^2 + dy^2 + dz^2$ and all positive signs, there would be no light cones. There would be no causality and no distinction between past and future. That minus sign in $ds^2 = -c^2dt^2 + dx^2 + dy^2 + dz^2$ is what creates everything that makes *time* feel like time, the arrow, the irreversibility, the difference between memory and prediction. We have seen that three spatial dimensions emerge from quantum entanglement structure, the pattern of correlations in a quantum network. But that explains static geometry. It does not explain evolution, causality, or the arrows. Time emerges differently. This is not due to the correlation pattern itself, but due to how the pattern changes. The difference lies not in which qubits are entangled, but in how entanglement spreads through the network. That spreading has a natural direction from less entangled to more entangled. The Lieb-Robinson velocity is its natural speed limit, the maximum rate at which correlations can propagate through a quantum network with local interactions. The light cone is the same as the entanglement cone. Causality determines the correlation structure. Time is a parameter that describes how this structure changes.

Then, why is time different from space? The minus sign before dt^2 is what fundamentally distinguishes time from spatial dimensions. For spatial dimensions, there is no distinction like this. The metric is always positive, given by $dx^2 + dy^2 + dz^2$. Every direction in space is the same. You can choose to move north, south, east, or west: the direction you pick does not matter. Time does not allow for such freedom. We move forward through time, not backwards. Our memories are of the past, not the future. Entropy always increases in one direction. Unlike time, space does not have this asymmetry. If time arises from quantum information, it should emerge in a different way from space.

Time emerges as a fresh insight when entanglement spreads through quantum networks. Start with a quantum system that is in an initial state, possibly with localised entanglement. Some qubits are entangled with their nearby neighbours, while distant regions remain uncorrelated. Let the system evolve under a Hamiltonian, quantum dynamics governed by Schrödinger's equation. What happens when this occurs? Entanglement spreads out. The correlations that are local in nature extend outward from there. Qubits that were not entangled initially become entangled now. Information then propagates through this network. This spreading of entanglement is time. More precisely, time is the parameter that describes how entanglement evolves. Spatial dimensions come from the pattern of correlations at a given instant, which shows how qubits are connected to each other. The temporal dimension comes from how this pattern changes, including how correlations spread and grow.

Time emerges through entanglement between a system and a clock, which is a stunning idea. In the Wheeler-DeWitt equation, the fundamental equation of quantum gravity, time disappears. The universe's wavefunction ψ\[g\] is dependent on the spatial metric g but contains no time parameter. The equation states:

$$\hat{H}\psi[g] = 0$$

where Ĥ is the Hamiltonian constraint. There is no time variable. The universe is described by a static wave function across the space of three geometries. This is known as the "problem of time". If the fundamental equation contains no time, then where does time originate? The *Page-Wootters mechanism* provides an answer. We have a system made up of two parts: S, the "system of interest", and C, the "clock". These two parts together form an entangled state:

|Ψ⟩ = Σ_t |ψ_S(t)⟩ ⊗ |t⟩_C (or ∫dt |ψ_S(t)⟩ ⊗ |t⟩_C for a continuous clock)

In this setup, the clock subsystem C is in a superposition of different time states, denoted by |t⟩. The state of the system, |ψ_S(t)⟩, is correlated with the time state of the clock. From the perspective of the total system, time is static. The wavefunction |Ψ⟩ remains unchanged. However, for an observer within the clock subsystem, or one that is entangled with it, time seems to flow. When you measure the clock and find it reading time t, the system is in state |ψ_S(t)⟩. If you measure again and find the time to be t', the system is then in state |ψ_S(t')⟩. Although the total wavefunction is timeless, you still experience evolution. Time is the correlation between a system and a clock. It emerges from entanglement, rather than from a fundamental temporal flow. This remains somewhat speculative: we lack a complete quantum gravity theory to fully validate it. But it suggests an elegant resolution, time emerges from entanglement between subsystems, with one subsystem serving as the clock.

Time from Thermodynamics

From thermodynamics, another perspective is the thermal time hypothesis, proposed by Carlo Rovelli and Alain Connes. *Consider* a quantum system in a mixed state ρ, which is not pure as it has entropy. The Hamiltonian does not define a preferred time direction, meaning all possible time evolutions are equivalent. But the state ρ defines a preferred time, which is thermal time, the parameter that describes the approach to equilibrium. From state ρ, we define the modular Hamiltonian:

$$K = -\log \rho$$

This operator generates time evolution:

$$U(\tau) = e^{(-iK\tau)}$$

where τ is thermal time. This is not the same as coordinate time in spacetime, but is intrinsic to the quantum state. For a thermal state at temperature T, thermal time is the same as ordinary time, but scaled by temperature. For other states, thermal time may be different, but it still gives us a notion of temporal flow. The thermal time hypothesis states that the time we experience is the thermal time of the macroscopic degrees of freedom with which we interact. We don't experience the fundamental quantum state, which may be timeless. Instead, we experience coarse-grained, thermodynamic variables, which have an intrinsic time direction: the direction of increasing entropy and the approach to equilibrium. Time emerges from thermodynamics. It is the parameter that describes systems as they evolve towards equilibrium.

The Arrow

Every theory of emergent time has to address a fundamental question: why does time flow forward, and what creates the arrow of time? In fundamental physics, time evolution is reversible. Schrödinger's equation is time-symmetric: if $\psi(t)$ evolves to $\psi(t+\Delta t)$, it can be reversed. Quantum mechanics treats the past and the future equally. Yet our experience is asymmetric. We remember the past, not the future. We see eggs break, but they never spontaneously reassemble. We age, decay, and die. There is a clear direction to time. The standard explanation is the second law of thermodynamics. Entropy increases. Systems evolve from order, which is low entropy, to disorder, which is high entropy. This creates what we know as time's arrow. But why does entropy increase? The second law is statistical, not absolute. It states that in an isolated system, entropy is likely to increase, but a decrease is not impossible; it is just very unlikely. The real puzzle is why the universe was in a low-entropy state initially. The universe started with unusually low entropy at the Big Bang, compared to its maximum possible entropy. This initial condition drives the increase in entropy that follows. But why have that initial condition? This leads to the past hypothesis—the assumption that the universe began in a special, low-entropy state. The reason for this is not explained by physics; it is simply a boundary condition. If time emerges from entanglement spreading, then the arrow of time has a natural explanation. Entanglement spreads naturally. When independent systems interact, entanglement grows. Information propagates and correlations deepen as a result. The process is effectively irreversible: unentangling a system is extraordinarily difficult. The arrow of time points towards increasing entanglement. We call this direction "forward" because it links to increasing entropy, which closely relates to entanglement in quantum systems. Time flows forward as entanglement spreads in the same direction. It is nearly impossible to reverse this process and unentangle a macroscopic system.

Then, what happens to causality if time emerges? Causality describes the events that can influence each other. Event A can cause event B if it lies in B's past light cone, meaning a signal can travel from A to B at or below light speed. In emergent spacetime, causality also emerges. What gives rise to it? It emerges from the way entanglement spreads through its structure. Information propagates through a quantum network at a finite speed, the Lieb-Robinson velocity. In a local Hamiltonian, qubits interact with their neighbours, and correlations spread at a maximum rate:

$$v\_LR \sim J \cdot a$$

Here, J represents the interaction strength and 'a' is the lattice spacing. This leads to an emergent speed limit, where information cannot travel faster than v_LR through the network. In emergent spacetime, v_LR corresponds to the speed of light. The spread of entanglement determines the causal structure, which relates to how certain events can influence others. Event A can influence event B if entanglement is able to propagate from A to B within the time available. This results in an emergent light cone.

The light cone structure, causality, and the distinction between spacelike and timelike all emerge from entanglement dynamics in the quantum network. Causality is not fundamental; it emerges from the propagation of quantum information.

Let us use the tensor network framework as the basis for this. A tensor network at a fixed "time" represents the spatial structure, showing which qubits entangle with which others. This captures the spatial geometry at a particular moment. Now *consider* dynamics. Apply a unitary operator U, which represents time evolution under a Hamiltonian. The tensor network then transforms:

$$|\psi(0)\rangle \rightarrow |\psi(t)\rangle = U(t)|\psi(0)\rangle$$

The evolved state has a different entanglement structure. Correlations have spread out. Qubits that were previously uncorrelated are now entangled. The tensor networks at t=0 and t=t are different. The sequence of tensor networks,

$$|\psi(0)\rangle, |\psi(\Delta t)\rangle, |\psi(2\Delta t)\rangle, \ldots$$

traces a path through quantum state space, which represents time. The "time direction" in hierarchical networks is orthogonal to spatial dimensions. It is the direction of increasing coarse-graining, where entanglement entropy grows and correlation lengthens. This is in line with the thermal time hypothesis, where time is the parameter describing how the state evolves under coarse-graining, known as RG flow, and it emerges from the entanglement structure and its transformation.

If time comes from entanglement, how do we explain time dilation? Yes. In emergent spacetime, curvature arises from entanglement patterns. Regions of different curvature have different entanglement structures. Time dilation emerges from entanglement dynamics - the rate at which entanglement spreads, or the "clock rate", depends on the local entanglement structure. In regions of high curvature, where gravity is strong, the network structure slows down the effective flow of time. Researchers have explored this phenomenon in AdS/CFT and tensor networks, where time dilation for clocks in different bulk positions equates to varying rates of renormalisation group flow for boundary operators across different energy scales. Deeper in the bulk, where the gravitational potential is higher, means that the clock rate is slower. This corresponds exactly to gravitational time dilation. Time dilation is not a fundamental property of time. It is a property of emergent time, which arises from how the entanglement structure affects the local rate of quantum evolution.

If time arises from entanglement, then what happened at the universe's beginning, the Big Bang? At the Planck time, 10^{-43} seconds after the Big Bang, the universe is very early and quantum gravity dominates. At this point, spacetime becomes uncertain and starts fluctuating. In the emergent picture, the Big Bang marks the beginning of entanglement spreading. Before this, the universe had minimal entanglement, perhaps just a single quantum degree of freedom with no spatial structure. Then entanglement started spreading. The quantum correlations grew, and the tensor network expanded. The three spatial dimensions we observe emerged. Entanglement spread further, and time itself emerged. The "initial state" wasn't in space or time. It was a quantum state of pure potential, with quantum information that had no structure. Space and time emerged together as entanglement spread. This is still speculative because we lack a complete theory of quantum gravity. But it is an appealing picture within the emergent spacetime framework. The Big Bang theory has a singularity, which is where classical general relativity breaks down. This predicts infinite density and curvature, but it is not a physical singularity. It is the boundary where emergent descriptions fail because the underlying quantum system is too simple and too unstructured for classical spacetime to be a good approximation. The concept of "before" the Big Bang isn't meaningful if time itself emerges. Since time begins with entanglement spreading, there is no "before".

If time emerges from entanglement, can it exist in a superposition? This sounds strange how can "when" be uncertain? But quantum mechanics makes this possible. *Consider* the quantum switch, an experiment where operations A and B are applied in different orders simultaneously. The system is in a state where "A then B" and "B then A" coexist. The temporal order is itself quantum. Experiments have tested this. Results agree with quantum predictions. The order of operations can exist in a superposition, creating interference effects between the two temporal sequences. If time emerges from entanglement, superpositions of temporal order are natural. The different branches of the wavefunction have different entanglement patterns, which correspond to different emergent timelines. From the perspective of the total quantum state, both temporal orders coexist together in superposition. Only when you measure something and become entangled with the system, determining the order, does the superposition collapse into a definite temporal sequence. Time is not fundamental. It's an emergent property of the entanglement structure. Like all emergent properties, it can be uncertain until it is measured, due to quantum effects.

Einstein's relativity suggests that the universe is a block: all of spacetime exists as a complete four-dimensional structure. The past and future coexist in this structure. Time, in this view, is not flowing; it is an illusion. But if time emerges from the spreading of entanglement, then the block universe picture is misleading. The past is a record of an entanglement that has already spread, forming quantum information that has decohered into classical correlations. The future is potential entanglement that has not yet spread. The present "now" is the interface between recorded entanglement, which is the past, and potential entanglement, which is the future. From this perspective, the past is a fixed record of correlations. The present is the current entanglement structure. But the future is not yet determined, and we will not realise it until entanglement spreads. This is more like a "growing block" universe than a static one. The block grows as entanglement spreads and quantum information changes into classical form, with the future becoming the past. But there's subtlety. In the Wheeler-DeWitt picture, the fundamental quantum state is a timeless wave function, and from this perspective, all of this evolution is static. The single wave function contains all moments. The terms "growth" and "flow" describe how subsystems interact with their environment. Time is relational, emergent, and observer-dependent.

The problem of time in quantum gravity is resolved when time is emergent. The Wheeler-DeWitt equation $\hat{H}\psi[g] = 0$ is timeless as it describes the fundamental quantum state, a state that is timeless. When we look at subsystems and their correlations, time emerges. The situation parallels how rigidity emerges in a solid. At the atomic level, there is no rigidity, only electromagnetic forces between nuclei and electron clouds. Rigidity is what you get when enough atoms lock into a lattice. It is not in the equations of quantum electrodynamics. It is in the collective behaviour. Time, in this picture, works the same way: it is not in the fundamental quantum state. It is in what happens when you zoom out and track how subsystems correlate with one another, considering only macroscopic variables. Similarly, the fundamental state of the universe is timeless. Time emerges when we look at subsystems and their correlations in a coarse-grained way. Different subsystems have different notions of time, with clocks ticking at different rates depending on their entanglement with the rest of the universe. This resolves several puzzles:

Why does the fundamental equation have no time? Because time is not fundamental. It emerges from correlations between subsystems.

Why does time flow? Because entanglement spreads, creating a natural direction from less entangled to more entangled.

How can a timeless equation describe an evolving universe? The evolution is in the correlations between subsystems, not in the overall state.

What is "now"? It is the current entanglement structure from the perspective of a particular subsystem, which in this case is you.

Time is not fundamental. It is an emergent property arising from quantum information dynamics. What this means for us is, when time is understood as an emergent phenomenon rather than a fundamental backdrop, several of physics' most enduring puzzles shift and in some cases dissolve entirely. The arrow of time loses its air of mystery: time moves in the direction of increasing entanglement and entropy because that is simply how quantum evolution unfolds when systems interact locally, with no additional explanation required. Time travel, by the same token, becomes not merely difficult but almost inconceivably so; reversing time would mean unentangling a macroscopic system, unwinding decoherence, and reversing entropy, a process that does not violate the laws of physics in principle but is overwhelmingly forbidden by the second law of thermodynamics in practice. The question of what happened at the beginning of time is reframed just as dramatically: if time itself emerges as entanglement spreads, then asking what came "before the Big Bang" may be as meaningless as asking what lies north of the North Pole; the question presupposes a temporal structure that did not yet exist. Time dilation, meanwhile, finds a natural home in this picture, arising because emergent time flows at different rates in regions with different entanglement structures; general relativity, in this view, captures the classical limit of that emergent flow. Quantum superpositions of time become not only conceivable but expected; different branches of the wave function can carry different temporal properties, and time remains genuinely quantum until the moment of measurement. Finally, causality itself emerges from the structure of correlation propagation: the light cone and the entanglement cone turn out to be one and the same, with the boundary of causal influence defined by how far entanglement can spread.

The emergent view complements physics and enhances our understanding. Features we considered fundamental are, in fact, consequences that arise from quantum information dynamics. Time is considered the fourth dimension, and it emerges in a different way than the three spatial dimensions. Spatial dimensions emerge from the entanglement structure, which includes which qubits are correlated, how strongly they are correlated, and the pattern of correlation. This is the architecture of the static geometry at a given moment. The temporal dimension emerges from entanglement dynamics: how correlations spread, how the pattern evolves, and how quantum information flows. This is the evolution by the change that occurs from one moment to the next. Structure is formed by space, and change is marked by time. Both are derived from quantum information. Neither is fundamental. Einstein unified space and time to form spacetime, a single geometric structure with four dimensions. Now we know that these four dimensions arise from quantum information and entanglement. The fourth dimension emerges alongside the first three. The source of this is not a static pattern, but pattern evolution. What matters is not which qubits are entangled, but how the entanglement spreads. Time exists, it flows, and it has direction, but it is not fundamental. This emerges from quantum information, like everything else.

13

Dynamics Without Time

Quantum mechanics has time. General relativity has time. But quantum gravity, the theory that combines them, does not. Well, it is not a defect but a fundamental requirement of the maths. In quantum mechanics, time is external. The Schrödinger equation,

ihbar∂ψ/∂t = Ĥψ, treats time t as a parameter that is independent of the quantum state ψ. Time flows, and the wavefunction evolves accordingly. In general relativity, time is internal. The metric $g_{\mu\nu}$ includes a time coordinate, which means time is part of the geometry being described, not external to it. When you combine these theories, specifically quantising general relativity, the features come into conflict. With time being a part of the quantum geometry, the quantum state has no external time parameter to evolve in relation to. The result is the Wheeler-DeWitt equation:

$$\hat{H}\psi[g] = 0$$

It is a constraint on the quantum state, not an evolution equation, and it has no time parameter or dynamics. This forces us to rethink emergence completely. Space emerges from the entanglement structure, which is what Chapters 8-11 have established. We described this emergence in terms of time evolution, but now we see that time itself must emerge. As a result, we need a framework that does not assume time exists from the start. The framework exists, and this chapter explains it. The Wheeler-DeWitt equation, derived in 1967 by John Wheeler and Bryce DeWitt, is the central equation of quantum cosmology. It describes the quantum state of the entire universe, $\hat{H}\psi[g] = 0$ Here, $\psi[g]$ is the wave function of the universe, g represents a three-dimensional spatial geometry (the metric on a spatial slice), and $\hat{H}$ is the Hamiltonian constraint from general relativity. This resembles quantum mechanics, with a Hamiltonian acting on a wave function, but it is different from Schrödinger's equation. Compare:

- Schrödinger equation ihbar$\partial\psi/\partial t = \hat{H}\psi$
- Wheeler-DeWitt equation $\hat{H}\psi = 0$

Schrödinger's equation includes time evolution, which is described on the left side by how ψ changes over time t. In contrast, the Wheeler-DeWitt equation has no time evolution; it is a constraint that states $\hat{H}$ acting on ψ equals zero. The wavefunction $\psi[g]$ assigns quantum amplitudes to different three-dimensional geometries. For each possible way space could be curved, $\psi[g]$ gives a complex number that represents the quantum amplitude for that particular geometry. But ψ doesn't change over time. The equation doesn't have a time parameter, so it tells us which quantum states are physically possible, not how they evolve. This is the "problem of time" in quantum gravity: the fundamental equation does not contain time. The timelessness of the Wheeler-DeWitt equation is not a flaw, but rather something forced by the structure of general relativity. In general relativity, the Hamiltonian constraint states that the total energy, including gravitational energy, is zero everywhere: $\hat{H}$_GR = 0. This is a constraint equation, not a dynamical equation. It says that at every point in spacetime, the energy density from matter and fields is exactly cancelled by the negative energy density from gravity. When you quantize this constraint, you get:

$\hat{H}\psi = 0$. The operator $\hat{H}$ acting on a quantum state should result in zero. This is analogous to the momentum constraint in electromagnetism, namely Gauss's law, which in quantum electrodynamics becomes an operator equation. But there is a sharp difference. In ordinary quantum field theory, fields are quantised on a fixed spacetime background. Time is external, it is part of the background, not something that is being quantised. In quantum gravity, spacetime itself is quantised. The geometry, including time, is a quantum degree of freedom. There is no external time parameter on which the wave function can evolve. This is why the Wheeler-DeWitt equation acts as a constraint, not an evolution equation, because there's no "outside time" for it to evolve with respect to.

The Space of All Geometries

To understand what ψ[g] describes, we need to *consider* the concept of superspace. Superspace is the space of all possible three-dimensional geometries. In superspace, each point represents one possible curvature of three-dimensional space. For instance,

Flat space has g_ij = δ_ij, which is the Euclidean metric.

The sphere has a g_ij corresponding to S^3 geometry.

Hyperbolic space has a g_ij corresponding to H^3 geometry.

An arbitrary curved space can have any positive definite metric g_ij.

Superspace is infinite-dimensional, meaning there are infinitely many ways to curve a three-dimensional space. The wavefunction ψ[g] is a function on superspace, assigning a complex amplitude to each point in superspace, which corresponds to each possible three-geometry. The Wheeler-DeWitt equation $\hat{H}\psi[g] = 0$ is a constraint on this function. The wave functions that satisfy this constraint are the only physically allowed quantum states of the universe. But there is no evolution in superspace. The wave-function ψ simply exists; it is a static function that assigns amplitudes to geometries. This concept is often referred to as the "frozen formalism" of quantum gravity. Fundamentally, there is no change. In superspace, everything exists at the same time.

How Dynamics Come Back

In the timeless wavefunction $|\Psi\rangle$, there is no evolution. But you can define conditional probabilities: given that the clock reads t_1, what is the probability that the system is in the state $|\varphi\rangle$?

$$P(\text{system in } |\varphi\rangle \mid \text{clock reads } t_1) = |\langle\varphi|\psi\_S(t_1)\rangle|^2$$

The probability is dependent on t_1. When you *consider* different clock values, the state of the system appears to change from your viewpoint.

We define the "effective state" as follows:

$$|\psi\_eff(t)\rangle \equiv (\text{system state} \mid \text{clock reads } t)$$

This effective state changes with the clock parameter t. It follows an effective Schrödinger equation:

$$ihbar\partial|\psi\_eff\rangle/\partial t = \hat{H}\_eff|\psi\_eff\rangle$$

where $\hat{H}$_eff is an effective Hamiltonian that is derived from the correlations present in $|\Psi\rangle$.

Dynamics emerge. You derive time-dependent evolution from a timeless total state by conditioning on a subsystem that acts as a clock. This resolves the problem of time: the fundamental description is timeless, but effective descriptions, which are conditioned on clock subsystems, exhibit time evolution. Which subsystem serves as the "clock"? In principle, any subsystem correlated with the others can function as a clock. Different choices yield different "time variables". Possibilities include the "clock" can be anything. A mechanical device like a pendulum or atomic oscillation, a scalar field whose value φ serves as time, a cosmological variable like the universe's scale factor, or even entanglement entropy, which increases as the system evolves.

Each choice gives a different time parameterization. The underlying physics, which are the correlations in $|\Psi\rangle$, stay the same, but the way it is described changes. For a harmonic oscillator, the oscillator's phase might be time. In cosmology, the universe's radius might be time. For a quantum field, the field's expectation value might be time. In emergent spacetime, the entanglement structure itself acts as the natural clock. When entanglement spreads and grows, it defines a temporal parameter. Regions with more entanglement are "later" than regions with less entanglement. The spread of correlations defines time.

Let's look at a specific example from quantum cosmology. Model the universe using just one degree of freedom, the scale factor a. This scale factor determines the size of the universe. The Wheeler-DeWitt equation then becomes:

$$[-\text{hbar}^2/(2m)\ \partial^2/\partial a^2 + V(a)]\psi(a) = 0$$

This looks like a time-independent Schrödinger equation where the universe acts as the "particle" and a represents its "position". Solve for ψ(a), which is the wave function that describes the quantum state of the universe as a function of its size. But there's no time. The wave function ψ(a) simply exists, assigning amplitudes to different universe sizes without changing over time. To extract dynamics, introduce a second degree of freedom: a matter field φ. The wave function becomes ψ(a, φ), which satisfies:

$$[-\text{hbar}^2/(2m\_a)\partial^2/\partial a^2 - \text{hbar}^2/(2m\_\varphi)\partial^2/\partial\varphi^2 + V(a,\varphi)]\psi(a,\varphi) = 0$$

Now, treat φ as the clock. For a fixed φ, the conditional state is:

$$\psi(a|\varphi) = \psi(a,\varphi) \ / \ [\int|\psi(a,\varphi)|^2 da]\char`\^(1/2)$$

As φ changes, the conditional state evolves. Define time as t ≡ φ, and you recover the effective dynamics for a(t). The universe appears to expand or contract as φ changes. The situation remains the same: the wavefunction ψ(a, φ) is timeless, and you are simply viewing it from different angles. In realistic scenarios, the Wheeler-DeWitt equation is complex. However, in the semiclassical limit, where quantum effects are small, the equation can be solved using the WKB approximation. Write the wave function as:

$$\psi[g] = \exp(iS[g]/\text{hbar})$$

where S[g] is a classical action. Substitute this into the Wheeler-DeWitt equation and expand in powers of hbar.

At leading order (hbar^0): $(\partial S/\partial g)^2 + V[g] = 0$

This is the Hamilton-Jacobi equation, which is the classical limit of quantum mechanics. The solutions S[g] define the classical trajectories in superspace, specifically the paths g(t) that the universe classically follows. The parameter t, labelling points along these trajectories, is what is known as emergent semiclassical time. This is not a fundamental concept, but rather it parameterises the classical path through superspace. In the WKB approximation, the wavefunction peaks along classical trajectories. Observers following such trajectories experience time as the parameter t. However, at a fundamental level, quantum mechanics implies that time does not exist, and the timeless wavefunction ψ\[g\] remains concentrated near classical paths. In modern cosmology, this semiclassical time is the same as cosmic time, which is the proper time measured by observers moving with the Hubble flow, who see the universe expanding uniformly around them. The radial coordinate z in AdS relates to the RG scale:

$$z \sim 1/\mu$$

A large z, which is deep in the bulk, corresponds to a small μ, indicating low energy in the IR. A small z, which is near the boundary, corresponds to a large μ, indicating high energy in the UV. The bulk also has a time dimension, which is an actual temporal coordinate, what does this correspond to when we reach the boundary? It corresponds to the evolution of the boundary state under RG flow. Different time slices in the bulk represent different stages in the evolution of the boundary under RG. The radial spatial dimension of the bulk directly maps to the energy scale of the boundary through the RG flow. In this context, time acts as a common constant. Our emergent model shows that the bulk's 'clock' is synchronised with the unitary evolution of the boundary's quantum state. Specifically, time emerges from the correlation between the network's sequential states, rather than the scale of the network itself. The temporal direction is the direction in which we move from a fine-grained to a coarse-grained description.

The Wheeler-DeWitt equation does not give you a film. It gives you every possible film simultaneously, superposed, with no preferred frame. Our universe, this one, with its particular galaxies and particular history; is one branch of a wavefunction that contains all possible universes. So why do we experience only one? This is where decoherence enters, and it is worth being honest that decoherence is not a complete resolution, it is the best one we have. When the universe interacts with itself, when the long-wavelength modes we can observe become entangled with the short-wavelength fluctuations we cannot, the superposition does not collapse. It merely becomes unobservable as interference. What remains, for any observer who cannot track every micro-degree of freedom, is the appearance of a single classical spacetime. We experience one history not because the others ceased to exist, but because we are entangled with a particular branch of the wavefunction and cannot see the rest. To an observer who does not track the environment, the system appears to be in a definite classical state, even though the total system, which includes the system and the environment, remains in a superposition. For the universe, what is the environment? The high-frequency modes are short-wavelength fluctuations that are below our measurement resolution, these are degrees of freedom we do not observe. When you coarse-grain the wavefunction, averaging over unobserved details, you get a definite classical geometry. The quantum superposition then decoheres into distinct classical branches. In the WKB approximation, the result is different classical trajectories. The wavefunction ψ = exp(iS/hbar) is largest along the classical paths where S has a stationary value, which are the paths of least action. Decoherence decides which classical trajectory becomes real. The universe follows this path not because it is special, but because decoherence stops quantum interference between different paths. Time moves along a decohered classical path, and we experience it because we are entangled with the coarse-grained and decohered degrees of freedom.

The Hartle-Hawking no-boundary proposal is a notable one in quantum cosmology. James Hartle and Stephen Hawking have proposed a theory that the universe's wave function can be found using a path integral over all possible compact four-geometries:

ψ[g] = ∫_{g_boundary = g} [Dg_interior] exp(iS[g_interior]/hbar)

First, fix the three-geometry g at the boundary. Then, sum over all the four-dimensional geometries that have the same boundary as g. This approach is timeless, with no evolution from a starting state. The method involves summing over all possible spacetime geometries at the same time. The result is a wavefunction ψ[g], which is a quantum state of the universe depending on its spatial geometry. This wavefunction satisfies the Wheeler-DeWitt equation. The question remains: what about the Big Bang? At what point does time begin? In the no-boundary proposal, there is no boundary, no initial moment. The geometry concludes in a gentle manner in the past, much like the South Pole on Earth. You can't ask "What's south of the South Pole?" because there's no answer. It's similar with "What existed before the Big Bang?", the question is meaningless. The geometry closes off, so time doesn't go beyond a certain point. In the region of this smooth initial geometry, quantum geometry dominates. There is no classical spacetime, no definite time. When you move away to larger scales, classical spacetime emerges, and the notion of time evolution emerges with it. A technical tool for quantum cosmology is the Euclidean signature. In quantum field theory path integrals, time can be expressed as imaginary values: $t \to -i\tau$. This process is called Wick rotation. The Minkowski metric:

$$ds^2 = -dt^2 + dx^2 + dy^2 + dz^2$$

becomes the Euclidean metric:

$$ds^2 = dx^2 + dy^2 + dz^2 + d\tau^2$$

Here, all signs are positive. Time is treated as "just another spatial direction." Why do this? Path integrals with a Euclidean signature converge more effectively due to their superior mathematical properties and simpler calculations. In the Hartle-Hawking proposal, integration is done over Euclidean geometries being smooth four-dimensional Riemannian manifolds, rather than Lorentzian ones. The Hartle-Hawking proposal puts forward a unique description of the universe's wave function, using a Euclidean path integral instead of a standard temporal path. Mathematically, this is expressed as $\Psi[g] = \int [Dg] \exp(-I\_E / \hbar)$, where I_E denotes the Euclidean action. By dropping the imaginary 'i' from the exponent, the geometry effectively smoothes out at the beginning of time, thereby eliminating the singularity and presenting a more persuasive argument for the origins of the universe. After computing the path integral in Euclidean signature, you analytically continue it back to Lorentzian signature, which gives you real-time and physical predictions. This approach treats time and space as equivalent in the path integral. They are not different but just distinct directions in geometry. The distinction emerges when you analyse the Lorentzian signature further. This supports emergent time, as in the Euclidean path integral, time behaves similarly to space. The difference emerges when you perform the analytic continuation to make physical predictions.

The Problem of Observables.

In quantum gravity, a deep puzzle is: what are the observables? In ordinary quantum mechanics, physicists use operators as observables, such as $\hat{H}$ (energy), $\hat{p}$ (momentum), and $\hat{x}$ (position), which they measure to obtain eigenvalues as results. In quantum gravity, geometry is dynamic. There is no fixed background spacetime to define observables against. This makes "position" ambiguous, what is it relative to? The coordinates are dynamic as well. In the Wheeler-DeWitt formalism, observables have to be diffeomorphism-invariant; they cannot depend on the choice of coordinates. This restricts what can be measured. One class of observables is relational quantities. Instead of asking "What's the field φ at position x and time t?" (which is not diffeomorphism-invariant), we ask "What's the value of φ when the scale factor is a?" (which is diffeomorphism-invariant). Relational observables do not refer to absolute positions or times, but instead describe the relationships between different physical quantities. This aligns with emergent time. If time is relational, defined by correlations between subsystems, then observables should be relational too. We should measure one subsystem's state relative to another, rather than relative to absolute time. An alternative framework for quantum mechanics without time: consistent histories. In standard quantum mechanics, you have a Hilbert space, a Hamiltonian, and time evolution:

$$|\psi(t)\rangle = \exp(-i\hat{H}t/hbar)|\psi(0)\rangle.$$

In consistent histories, the time evolution is eliminated. Instead, histories are defined and these are sequences of events or properties at different "times". A history is a chain:

$$(A_1 \text{ at } t_1) \rightarrow (A_2 \text{ at } t_2) \rightarrow (A_3 \text{ at } t_3) \rightarrow \dots$$

where each A_i is a proposition, a projection operator, and t_i are time labels. To assign probabilities to histories, you use the decoherence functional:

$$D(H,H') = Tr(P\_H \; \rho \; P\_\{H'\})$$

where P_H is the projection operator for history H, and ρ is the initial density matrix. If histories decohere - $D(H,H') \approx 0$ for $H \neq H'$ - you can assign consistent probabilities without considering time evolution. The time labels t_i are parameters that lack fundamental significance. Changing the time parametrization gives you a different but equivalent description. This formulation fits naturally with quantum gravity. The Wheeler-DeWitt equation yields a timeless state $\psi[g]$. Histories consist of sequences of geometries. The system uses the decoherence functional to assign probabilities. Consistency is checked. In consistent histories, time emerges as a label for events, and it is not fundamental to the formulation.

We observe classical spacetime to have a definite time evolution for practical purposes. How do we get from the timeless Wheeler-DeWitt framework to this classical picture? The transition occurs through the semiclassical limit and decoherence. Consider the early universe. At Planck-scale energies (10^{19} GeV), quantum fluctuations in geometry are dominant. Spacetime is quantum and uncertain, so classical notions of time do not apply. As the universe expands and cools, quantum fluctuations lose their coherence. High-frequency modes become entangled with low-frequency modes, which cause the wavefunction to split into separate classical histories. In each branch, a classical spacetime emerges. This spacetime has a metric $g\mu\nu(x)$ that includes a time coordinate t. This time coordinate is what we refer to as the emergent time, and it labels the classical trajectory of the branch. Observers, like us, measure time with physical clocks, such as atoms and oscillators, that track the emergent time coordinate. We experience time as flowing due to our entanglement with the decohered classical geometry that surrounds us. The fundamental theory, Wheeler-DeWitt, is timeless. The effective theory, classical general relativity, has time. Time emerges when we transition from quantum to classical. How do dynamics emerge from a state of timelessness? There are multiple complementary perspectives. The concept of emergent time is supported by multiple frameworks, which were converging on similar conclusions. In the relational time framework, time is defined as the correlation between subsystems, and its "flow" is relative to the system being measured by a clock. The conditional evolution approach extracts a Schrödinger-like equation from the Wheeler-DeWitt equation by using a clock subsystem, which has the effect of recovering familiar quantum dynamics. The WKB approach also demonstrates the emergence of time, but only when gravity is treated nearly classically, and quantum fields evolve on a geometric background that is changing slowly. Entanglement growth along the RG flow provides a natural direction from the information-theoretic perspective, and MERA layers in tensor networks represent different scales that can be viewed as time steps in the bulk. Decoherence plays a complementary role by selecting a preferred basis through environmental interactions, which breaks the timelessness of the quantum state into a sequence of effectively classical moments. The consistent histories framework further supports this idea, where time labels events within decoherent histories, allowing different histories to define different time orderings, with consistency conditions choosing those that align with observation.

These perspectives converge on one insight: time is not fundamental. Dynamics emerge from correlations, coarse-graining, and decoherence that occur within a timeless quantum state. The Wheeler-DeWitt equation describes a universe without time. However, when you *consider* subsystems, observers, semiclassical limits, or consistent histories, time emerges from correlations between subsystems, just as space emerges from entanglement between them. Recognising that dynamics arise out of timelessness helps to resolve several puzzles:

The problem of time in quantum gravity. Time is not fundamental, it emerges. The Wheeler-DeWitt equation is timeless because the fundamental description goes beyond time. Time appears when we use semiclassical perspectives or focus on subsystems. Before the Big Bang, this question may be meaningless. Time emerges when spacetime becomes classical. As a result, there's no "before" the Big Bang, because time does not exist beyond the point at which classical spacetime breaks down. The arrow of time, arises from decoherence and the increasing entanglement. The direction of time corresponds to an increase in entropy and coarse graining. Quantum superpositions of time are possible in the fundamental description. Different branches can have different temporal structures, and observers in these branches experience time differently. The nature of change that is change is not fundamental. It occurs when subsystems are associated with different clock values. Although the total state itself remains unchanged, observers within that state see everything as evolving.

This radically alters our conception of time. It is an emergent feature, not a fixed reality, and it becomes apparent when we *consider* the quantum state from our perspective. Time is real, it flows, and it matters. But time is not fundamental: it emerges from quantum information, from correlations, from coarse-graining, and from the structure of the universe's timeless quantum state. Dynamics without time are not just possible; they are the nature of reality at its deepest level. The Wheeler-DeWitt equation governs the quantum state of the entire universe, and its most exciting feature is that its eigenvalue is zero. Nothing changes. The wave function is frozen, timeless, a static superposition of all possible configurations. And yet here we are, experiencing time. The resolution is not that the equation is wrong. It is that time is not what we thought it was.

14

Five Paths to One Answer

Physicists rarely agree on anything fundamental. If you ask ten theorists about quantum gravity, you will get fourteen different opinions. Across the domain of quantum gravity research, which includes string theory, loop quantum gravity, causal dynamical triangulations, asymptotic safety, and causal sets, one conclusion consistently emerges. Spacetime, it turns out, is emergent. Five frameworks built in opposition to each other, by communities that have barely spoken for decades, arriving at the same conclusion. That is not interesting. That is a signal. These approaches begin with completely different premises. String theory quantises one-dimensional objects in a background spacetime. Loop quantum gravity, in contrast, quantises spacetime geometry directly. Causal dynamical triangulations build spacetime using discrete building blocks. Asymptotic safety, meanwhile, treats gravity as a quantum field theory. They use different maths. They are based on different assumptions. They were developed by different communities, often in isolation or opposition to each other. Yet they all arrive at the same place: spacetime emerges from something more fundamental. What convinces me is that different approaches agree on this idea. One approach could be wrong, like string theory about extra dimensions or loop quantum gravity about spin networks. But five different frameworks, made by separate groups using different maths and having different tastes, all say spacetime is emergent.

This is not a coincidence. It shows the idea is real. Dimensions are not the basic inputs; they are outputs that are derived. And this "something deeper" always seems to involve quantum information. We will now read what each approach tells us about emergent spacetime and why it matters that they converge. If different paths all lead to the same place, that place is probably real. String theory replaces point particles with one-dimensional extended objects, called strings. The different vibration modes of a string produce different types of particles, including electrons, photons, quarks, and gravitons. For mathematical consistency, string theory needs extra dimensions.

Bosonic string theory requires 26 dimensions; superstring theory, which includes fermions, needs 10; and M-theory, which unifies the different superstring formulations, exists in 11.

We can only observe four dimensions: three are spatial and one is temporal. So, where are the extra dimensions? String theory needs ten dimensions to be mathematically consistent. We observe four. The theory's answer to this discrepancy is compactification, the six missing dimensions are curled up so tightly that they fit inside a space roughly the size of the Planck length, 10^{-35} metres, ten million billion times smaller than what the Large Hadron Collider can probe. They are not hidden in the sense that someone put them there. They are hidden the way the circumference of a garden hose is hidden from someone viewing it from a kilometre away: real, but too small to resolve. The shape of those curled dimensions determines the physics we observe in the large ones. Different shapes give different particle masses, different coupling constants, different versions of the Standard Model. This is where the string theory landscape comes from the 10^{500} possible vacuum states, each corresponding to a different way of folding six dimensions into a Planck-scale geometry. I raise this not to endorse string theory but because it poses a question that cuts to the centre of this book: if the number of large dimensions is a consequence of how extra dimensions compactify, is dimensionality a fixed feature of the theory, or an output? And if it is an output, what selects three? The shape of the hidden dimensions affects the physics in the larger ones. When different methods compactify these dimensions, they produce different masses of particles, forces, and coupling constants. But the main point for our purposes is this: in string theory, the familiar four dimensions are also emergent. Strings propagate in spacetime, but spacetime itself is a result of string dynamics. The theory is formulated from abstract string configurations, known as the string field, at a fundamental level. When we study the propagation of strings, spacetime emerges as a concept. Matrix theory shows this very clearly. It's a way of formulating M-theory where spacetime comes from how matrices behave - these are infinite-dimensional quantum operators. The matrices themselves aren't part of spacetime, but are more like abstract algebraic objects. When we study how they behave, we can find out about the geometry of spacetime. The spatial dimensions, all of them, including our three large ones, emerge from the eigenvalue distributions of these matrices. The matrix dynamics change the emergent geometry. String theory's specific contribution to our question is not just that spacetime is emergent, it is that AdS/CFT makes this precise in a way no other approach has yet matched. The correspondence is exact, not approximate. One dimension of space emerges from an energy scale. The geometry is computable. If string theory is wrong about everything else, it has still given physics the most explicit demonstration available of how a spatial dimension can be a derivative rather than a fundamental quantity. The question of whether string theory is the right theory of quantum gravity is separate from the question of whether the lesson it has shown us is real. And I think the lesson is real.

A Different Route

Loop quantum gravity (LQG) uses a different method: it quantises general relativity directly, without assuming there is a background spacetime. The development of loop quantum gravity is attributed to the Indian physicist Abhay Ashtekar, who was born in Kolhapur, Maharashtra. Ashtekar's reformulation of Einstein's equations in 1986, using a new set of variables, was an astonishing breakthrough. These variables, now called Ashtekar variables, made it possible to express general relativity in the language of gauge theory, and paved the way for the application of quantum field theory to gravity. The work of Carlo Rovelli and Lee Smolin, who have built on Ashtekar's reformulation, has led to the development of the spin network picture, which is central to the field of loop quantum gravity today. All the major results in loop quantum gravity, including area quantisation and the counting of black hole entropy, have their origins in Ashtekar's pioneering work in 1986. In LQG, spacetime geometry is represented by spin networks, graphs where edges have quantum numbers, known as spins, and nodes represent units of quantum geometry. A spin network is a graph whose nodes are junction points and whose edges carry SU(2) spin labels (j = 1/2, 1, 3/2, ...). Each edge contributes to geometric quantities like area and volume, the eigenvalues of the corresponding operators are discrete, which is why area and volume are quantised in this framework.

The result is that geometry is quantised. Area and volume come in discrete multiples of the Planck area (10^{-70} m^2) and the Planck volume (10^{-105} m^3). At the Planck scale, spacetime isn't smooth and it's made up of a discrete network of quantum geometric units, known as "spin foam," which connect three spatial dimensions with one time dimension.

So does spacetime emerge in LQG? Yes, it emerges in two ways: *First*, the discrete spin network averages to a continuous geometry at scales that are much larger than the Planck length. This is where the metric g_μν emerges as a coarse-grained description of the underlying quantum structure. Think of fluid dynamics from the perspective of molecules. At a molecular level, water consists of discrete molecules in motion. At larger, macroscopic scales, it behaves like a continuous fluid that can be described by the Navier-Stokes equations. The continuum emerges when you average out the microscopic details. In a similar way, smooth spacetime is obtained by averaging the Planck-scale spin networks.

Secondly, the spin network is quantum in nature. This means the full quantum state consists of a superposition of different networks:

$$|\Psi\rangle = \Sigma\{networks\}\ c\{\text{network}\}\ |\text{network}\rangle$$

As the superposition decoheres through interaction with matter, classical geometry emerges. One particular network, or a small group of networks, becomes dominant. In LQG, the continuum structure and classical geometry are both emergent. The description is quantum, discrete, and geometric. We observe geometry emerging from quantum superpositions and coarse-graining. One of the achievements of LQG is the derivation of black hole entropy. In this framework, the event horizon is a surface made up of Planck-area quanta. Each quantum can have different states, such as different spins. The number of ways to arrange these quanta, or the number of microstates, gives the entropy. By counting microstates, LQG derives a formula proportional to the Bekenstein-Hawking result:

$$S = A/(4l\_P^2)$$

(This requires fixing the Barbero-Immirzi parameter $\gamma \approx 0.274$ to match the Bekenstein-Hawking coefficient, a free parameter LQG fixes empirically rather than predicting from first principles.) What is striking is that LQG, which takes a completely different approach from string theory, arrives at the same entropy formula. It also confirms that entropy is proportional to area, not volume. This area law, the signature of holography, appears in both string theory and LQG. It tells us something deep about spacetime: information lives on surfaces, not in volumes. The geometry has fewer degrees of freedom than naive volume counting would suggest. This supports emergent dimensionality. If three-dimensional bulk geometry were fundamental, we would expect entropy to scale with volume. Instead, it scales with area, suggesting that one dimension is "redundant," emergent, and holographically encoded.

Other Approaches

Causal dynamical triangulations (CDT) takes a different kind of courage. It throws away the smooth manifold entirely. Instead of assuming spacetime is continuous, it builds it from pieces, triangles in two dimensions, tetrahedra in three, 4-simplexes in four and then does what quantum mechanics demands: it sums over every possible way to assemble them. Not every configuration. All configurations simultaneously, weighted by their quantum amplitudes. The question is whether, when you take the limit as the pieces get infinitely small, something smooth emerges, whether the universe reassembles itself from its own rubble. In CDT, it does. But only with one condition causality must be respected. The pieces are not allowed to glue together in ways that create closed time loops. Time is given a direction before the sum is done, and the geometry that comes back has three spatial dimensions and one temporal dimension, at large scales. At small scales, near the Planck length, something stranger happens. The "causal" aspect is the hinge: CDT requires that simplexes be glued together in a way that preserves time ordering. No closed timelike curves or causal pathologies are allowed. What does CDT reveal about emergent spacetime? CDT simulations show surprising behaviour of dimensionality. At large scales, spacetime has three dimensions of space and one of time. However, at the Planck length, its effective dimensionality reduces to about two. This comes from measuring the *spectral dimension*, which tells us how distances scale in the network. This in ordinary d-dimensional space is equal to d. At large scales, CDT recovers four dimensions, three of space and one of time, the universe we live in. But as you probe smaller and smaller distances, approaching the Planck length of 10^{-35} metres, the spectral dimension, the way distances actually scale in the network, begins to fall. By the time you reach Planck scale, it has dropped to roughly two. The geometry shrinks. This is not a quirk of CDT. Loop quantum gravity finds the same thing. So does asymptotic safety. Three independent approaches to quantum gravity, built on entirely different assumptions, all agree: at short distances, spacetime has fewer effective dimensions. The universe, at its deepest level, may be flatter than it looks. This suggests that dimensionality is emergent. Our familiar four dimensions appear only at large scales. At a fundamental level, spacetime may have fewer dimensions, or it may not be well defined at all. CDT also shows phase transitions, which are abrupt changes in geometry when parameters like the cosmological constant change. The geometry is unrecognisable in some phases, but in others, a four-dimensional spacetime appears. The fact that geometry can be "turned on" or "turned off" shows it is not fundamental. It is an emergent feature of the underlying quantum system.

Asymptotic safety suggests an intriguing possibility; quantum gravity can be a renormalisable quantum field theory in four dimensions, but it is renormalisable in a non-perturbative way rather than a perturbative one. The idea is that general relativity, when quantised as a field theory, has a non-trivial UV fixed point. As you go to higher energies, or shorter distances, the theory flows to this fixed point, where the coupling constants stabilise. The theory then becomes well-defined at all scales. If asymptotic safety is effective, then string theory and extra dimensions are not required. Quantum gravity can be seen as quantum general relativity that has been properly formulated. What does asymptotic safety reveal about emergent spacetime? At the UV fixed points at Planck-scale energies, scale invariance holds. The geometry loses its classical meaning because the metric fluctuates wildly. There is no classical spacetime. As you move away from the fixed point towards lower energies, a smooth geometry emerges. The metric fluctuations decrease, and classical spacetime starts to take shape. Dimensionality itself may change. The effective dimension is different from four. Studies suggest that there is a reduction in spectral dimension like CDT, with fewer dimensions at high energies. In asymptotic safety, spacetime emerges from the RG flow as it moves away from the UV fixed point. At the fixed point, classical geometry is absent. As you flow towards the IR, geometry emerges. Time emerges as the RG flow parameter, which is the direction away from the UV fixed point towards the IR. Spatial dimensions emerge as the theory crystallises, breaking scale invariance.

Causal set theory makes the most minimal bet of any approach that is all that exists is events, and the causal relations between them. That is it. No continuous manifold. No metric. No coordinates. Just a discrete collection of happenings, partially ordered by which can influence which (if $e_1 \prec e_2$, then e_1 can causally influence e_2). The geometry distances, angles, dimensionality; comes later, derived from the pattern of relations rather than assumed from the start. It is, in a sense, the physicist's version of what the philosopher Gottfried Leibniz argued three centuries ago: space is not a container that things exist in. Space is the set of relations between things.

The causal set theorist is simply trying to build that idea into equations that survive contact with quantum mechanics.

The causal structure, which determines what events can influence others, is primary. This derives the geometry. How? This is done by analysing the pattern of causal connections. In a causal set that approximates smooth spacetime, the number of elements in a point's causal future determines its four-volume. The structure of maximal chains, which are the longest sequences of causally linked events, encodes dimensionality. Causal set theory enables the emergence of smooth spacetime in a continuum limit, where discrete events are densely packed and effectively form a continuous manifold. Emergent dimensionality also appears. The causal structure extracts the effective dimension, rather than it being an input. Different causal sets have different emergent dimensions. Recent studies in causal set theory suggest that dimensional reduction occurs at small scales, similar to CDT. The effective dimensionality may depend on the scale at which you probe the causal structure. Causal sets embody emergent spacetime in this way. The fundamental ingredients are discrete events and the causal relations between them. These relations give rise to geometry, dimensionality, and metrics.

An innovative proposal is that gravity itself emerges from thermodynamics. Erik Verlinde proposed in 2010 that gravity is not a fundamental force but an entropic force, similar to the force from a compressed spring or a stretched rubber band. These forces arise from the statistical tendency of systems to maximise entropy. Verlinde's argument is that spacetime has entropy associated with horizons, such as black hole horizons and Rindler horizons. When matter is present, moving it alters the entropy. The entropic force that pushes matter towards maximum entropy is what we refer to as gravity. Apparently, this entropic force follows Einstein's equations. Gravity is derived from thermodynamics and holography. A notable demonstration of this concept came from the work of Shiraz Minwalla at the Tata Institute of Fundamental Research in Mumbai. The team, including Minwalla, made a significant discovery in 2007, which revealed that the equations governing black holes in

Anti-de Sitter space have a precise mathematical correlation with the Navier-Stokes equations of fluid mechanics. Although these two sets of equations appear to describe vastly different phenomena. One relates to spacetime curvature and the other to the flow of water in a pipe; the holographic dictionary can provide accurate translations between them. This discovery suggests the idea that spacetime geometry is just one of several equivalent languages that can describe quantum information on a boundary, and not the only valid description. If Verlinde is correct, spacetime is emergent in itself. Temperature arises from the motion of molecules, and similarly, spacetime might arise from the microscopic degrees of freedom that contribute to entropy. What are these microscopic degrees of freedom? Verlinde does not specify them definitively, but quantum entanglement and quantum information are likely candidates. The entropy in question is entanglement entropy, which drives emergent dynamics. This idea is connected to our theme. Gravity emerges from entropy, and entropy comes from entanglement. Since spacetime geometry is closely linked to gravity in general relativity, it also emerges from entanglement. Emergent gravity remains controversial due to some unresolved issues. However, it is in line with the growing consensus that spacetime is not fundamental. Thanu Padmanabhan, a renowned Indian theoretical physicist at IUCAA in Pune, has confirmed the same conclusion independently, though by a different method. Over twenty years of extensive research have shown that Einstein's field equations and a broad class of generalisations can be derived systematically from a single thermodynamic identity governing horizons. Padmanabhan's work shows that gravity is a natural consequence of applying entropy's mathematical framework to surfaces that impede information flow. His research also suggests the universe tends towards holographic equipartition, where surface degrees of freedom balance those in the enclosed volume. This finding, reached through thermodynamic reasoning without relying on AdS/CFT or entanglement, supports the holographic accounting notion we have been exploring. That thermodynamics, holography, and quantum entanglement all converge on the same conclusion, a sign that these approaches illuminate the same fundamental truth. Learn from the different methods used here. Spacetime emerges from the dynamics of strings, branes, and matrices in string theory. The dimensions, including the large ones, are not fundamental, they emerge from how fundamental objects move. In loop quantum gravity, a smooth geometry emerges from discrete quantum spin networks at large scales. The quantisation of area and black hole entropy supports the idea of holography, and the number of dimensions may be reduced at the Planck scale. Causal dynamical triangulations shows four-dimensional spacetime emerging from simple discrete units at large scales, where the spectral dimension drops to around 2 at the Planck scale and the geometry undergoes changes in phase. In asymptotic safety, spacetime emerges from the flow of the renormalisation group away from a fixed point in the ultraviolet range, at which classical geometry does not exist and the effective dimension changes with scale. Causal sets take a different route, where the geometry and dimensionality emerge from a discrete causal structure, smooth spacetime is the limit when this structure is continuous, and the effective dimension depends on the scale. Finally, in emergent gravity, gravity and spacetime emerge from thermodynamics and the maximization of entropy, where the fundamental elements are the microscopic degrees of freedom, possibly entanglement. The starting points differ; strings, spin networks, simplexes, renormalisation flows, causal sets, entropic forces. The conclusion does not. In every case, spacetime dissolves into something more fundamental when examined closely enough.

What About Dimensionality? These approaches partially explain why we observe three spatial dimensions. The observations are:

- Dimensional reduction at high energy. CDT, asymptotic safety, and some causal set models show that the effective dimensionality decreases at small scales. Perhaps the fundamental description has fewer dimensions, possibly two, and we observe four dimensions at larger scales. This is related to holography, as it explains why we see (3+1)-dimensional spacetime if the basic theory is (1+1)-dimensional or (2+1)-dimensional and extra dimensions come from holography.

- Stability of four dimensions. In string theory, compactifications to four large dimensions are more stable than others. Compactifications with flux that stabilise extra dimensions usually result in four large dimensions. Why? This is partly due to the balance discussed in Chapters 10-11, where three spatial dimensions optimise information capacity, computational complexity, and stability. String theory discovers this balance through its dynamics.
- Phase structure. CDT indicates that a geometry with four dimensions is a stable phase. The other dimensionalities are associated with different phases, and four dimensions are the preferred geometry across a wide parameter range.
- Quantum error correction. If spacetime is a quantum error-correcting code, as suggested by AdS/CFT and holography, then the dimensionality reflects the code's structure. The three spatial dimensions may be optimal for certain coding properties.

No single approach fully explains "why three" from first principles, except for the arguments presented in Chapters 10-11. However, various approaches converge on that four-dimensional spacetime, which consists of three spatial dimensions and one time dimension, is special. It is stable, natural, and optimal. Quantum information is central to all approaches. In string theory, strings and branes are quantum objects. Their interactions give rise to entanglement. The geometry of spacetime that emerges is a pattern of entanglement. In loop quantum gravity, spin networks are quantum states. They exist in a quantum superposition, which is quantum information. The entropy of a black hole counts its quantum microstates, revealing the role of quantum information. In causal dynamical triangulations, simplexes serve as classical building blocks. However, the sum over geometries represents a quantum superposition, where the emerging geometry is determined by quantum amplitudes, and quantum information dictates the spacetime. In asymptotic safety, the RG flow describes how quantum correlations change with scale. Coarse-graining means a loss of information, and the emergent geometry shows which information remains. In causal sets, discrete events can be quantum in nature. Their causal structure may exist in a state of superposition, with quantum information deciding which causal paths contribute to the emergent spacetime. Entanglement entropy, a type of information, is the main factor in emergent gravity, driving the emergence of geometry. Quantum information is the common language. Look closely enough into any approach, and you will find that quantum information emerges as the fundamental substrate. The way quantum information is organised, entangled, and evolves determines spacetime.

This reinforces our argument from Chapters 8-12, that three dimensions emerge from the entanglement structure. It doesn't just apply to one approach to quantum gravity; it's a universal principle applying to all approaches. If spacetime is derived from quantum correlations arranged in their most stable configuration three spatial dimensions should emerge naturally. This is true regardless of whether the theory is based on strings, loops, simplexes, or something completely different.

It is an intriguing possibility that different quantum gravity approaches describe the same fundamental reality from different viewpoints. *Consider* string theory and loop quantum gravity, which appear to be contradictory. This contradiction is between strings and loops, extra dimensions and four dimensions, and continuous and discrete theories. Yet, if both theories effectively describe emergent spacetime, they could be dual. String theory may be the right description when the string coupling constant g_s is small, which is known as weak coupling, and strings behave like separate objects. At strong coupling, loop quantum gravity might be the right description, where strings come together to form a network, a spin network. Or both might be semiclassical limits of a deeper theory that is formulated entirely in the language of quantum information, without strings or loops.

The hints are suggestive. Both approaches reproduce the same black hole entropy formula, $S = A/4$. Both imply Planck-scale discreteness, though through different mechanisms. And both exhibit dimensional reduction at high energies, where the effective dimensionality drops from four toward two. Unifying them might need abandoning spacetime in the fundamental description. We begin with quantum information, including abstract qubits, entanglement, and quantum circuits, from which spacetime can emerge as an effective description. Different effective descriptions, such as strings, loops, and simplexes, can appear in different limits or phases. This could be the ultimate emergence: spacetime and all quantum gravity emerging from quantum information. We're not there yet, but the convergence of approaches suggests we're on the right track.

Lessons for Three Dimensions

Quantum gravity lessons reveal something about our main question: why are there three spatial dimensions?

- Lesson 1 Dimensionality is not fundamental. All approaches indicate that dimensions can change with scale, emerge in phase transitions, or come from more basic structures. The number of dimensions, such as four, should be seen as an outcome rather than a starting point.
- Lesson 2 Holography constrains dimensions. The area law for entropy consistently appears across different approaches. This implies a holographic structure, where bulk information is encoded on boundaries. Holography can produce large dimensions from fundamentals that are (d-1)-dimensional.
- Lesson 3 Information-theoretic optimisation matters. Quantum information theory explicitly appears in areas like AdS/CFT, emergent gravity, and quantum error correction, where dimensionality emerges from optimising capacity, complexity, and stability.
- Lesson 4 Three plus one is special. Four-dimensional spacetime emerges naturally from various sources, such as string theory compactifications, CDT phase structure, or asymptotic safety fixed points. This supports our argument that three spatial dimensions are optimal.
- Lesson 5 Quantum information is universal. Across all approaches, entanglement and quantum information are indispensable. This supports

our argument that dimensionality emerges from the entanglement structure.

The quantum gravity community is discovering what we have argued throughout, as they explore different paths: spacetime is emergent, dimensions are not fundamental, and three spatial dimensions arise naturally from quantum information dynamics. We still don't have a complete theory of quantum gravity. String theory is still incomplete. Loop quantum gravity has some unresolved issues. Causal dynamical triangulations are still being developed. Asymptotic safety has not been proven yet. Causal sets remain incomplete. The convergence is unmistakable, and all approaches state that spacetime is emergent, with its foundation in quantum information. The next step is to make this precise. We require a model of quantum gravity, possibly a simplified one, that is completely expressed in the language of quantum information. We have to show in detail how spacetime emerges and how the three spatial dimensions arise naturally. AdS/CFT is our closest example, which precisely shows how spacetime emerges from the quantum information in the CFT state. However, AdS isn't our universe because we do not live in negatively curved space. Extending AdS/CFT to de Sitter space, which is our accelerating universe, or to flat space, is ongoing work. If this work is successful, we will have a complete picture of our universe, with its three spatial dimensions and time, emerging from quantum information. Or, tensor network approaches could have the answer. By creating explicit networks of qubits and entanglement, we may be able to show that three spatial dimensions emerge, as per the logic outlined in Chapters 10-11. The mathematics are challenging, but the vision is clear. Spacetime and dimensions emerge. Quantum gravity confirms the optimal configuration of three spatial dimensions when approached from multiple directions.

V

The Lab Report

These ideas are no longer confined to blackboards.
Quantum simulators are building toy universes.
Gravitational wave detectors are listening for the seams.

15

Spacetime on a Chip

For the first time, physicists demonstrated the mathematics of holographic spacetime on a quantum simulator in 2019. A team of researchers created a perfect tensor state, the fundamental building block of holographic codes, using a six-qubit nuclear magnetic resonance processor, as proposed by Pastawski, Yoshida, Harlow, and Preskill. Measuring entanglement entropies across different partitions of the qubit register, they obtained results that matched the predictions from the Ryu-Takayanagi formula, which relates entanglement to geometric area in AdS/CFT. Although the system was small, consisting of just six qubits that encoded a single tensor rather than a full holographic network, the principle was clearly established. Researchers verified the relationship between entanglement and geometry on actual hardware, achieving a landmark breakthrough that goes beyond theoretical confirmation. Distance, in this setting, is not measured in metres but read from entanglement strength. Area is not calculated from coordinates but counted in entanglement bonds. Curvature arises from the way correlations arrange themselves across the network. This is the experimental frontier of emergent spacetime, where quantum simulators are used to test whether geometry, dimensions, and causality emerge from entanglement as theory predicts. Currently, we do not have the capability to build three-dimensional spacetime with gravity in a laboratory setting, as this would necessitate a larger number of qubits than we have. We can create simplified models and demonstrations that show how geometry is formed from quantum information. These models and demonstrations help us understand the process, and this chapter explains the methods we are using to achieve this. That was 2019. The field has accelerated dramatically since then. Google unveiled Willow, a quantum processor, in December 2024, which crossed a threshold that physicists had been pursuing for nearly three decades. As an engineering student, physics becomes personal for me here. Since these are not just theoretical experiments, but actual machines that somebody has carefully designed, fabricated, and programmed. Somebody has laid out the qubit layout, somebody has built the chip, and somebody has written the calibration code. In my lifetime, the concept of "entanglement builds geometry" has transitioned from a theoretical idea to a laboratory result, and this transition is still ongoing. The chip's 105 qubits are notable, but the breakthrough comes from scaling up the technology, which holds greater promise. For the first time, the addition of more qubits resulted in a decrease in errors, rather than an increase. As the system scaled up from a distance-three surface code to distance-five and then distance-seven, the error rates decreased exponentially, with each incremental increase in code distance cutting the error rate by approximately half. The team have a major breakthrough with this achievement, which showcases the real-world potential of quantum error correction. What makes this achievement so big is that it shows larger, more reliable quantum computers can be built by scaling up, rather than just improving what already exists.

The same transition appeared elsewhere. By the end of 2023, the thousand-qubit barrier was broken by Atom Computing, which showcased a quantum system with an atomic array of twelve hundred and twenty-five sites, eleven hundred and eighty of them being populated with neutral ytterbium atoms that had trapped in optical tweezers, and were held still by laser beams. The coherence times reached forty seconds, an eternity in quantum mechanics where states typically collapse within microseconds. The atoms maintained their quantum information long enough to perform complex operations, error-correct mid-computation, and prepare entangled states impossible in smaller systems.

By November 2025, IBM had unveiled its Nighthawk processor, featuring one hundred and twenty qubits connected through two hundred and eighteen tunable couplers in a square lattice configuration. The architecture of the Nighthawk processor enables the creation of complex circuits with up to five thousand two-qubit gates, which are essential for establishing quantum correlations and advancing quantum computing capabilities. IBM's roadmap targets seventy-five hundred gates by 2026, ten thousand gates by 2027, and fifteen thousand gates by 2028. These milestones are major advances, not minor upgrades. They signal a major shift from noisy intermediate-scale quantum computers to more reliable systems that can detect and correct errors at a faster rate than new errors occur. Denmark's Export and Investment Fund and the Novo Nordisk Foundation have a major investment of eighty million euros in QuNorth, a Nordic quantum initiative, which they announced in mid-2025. The system, named Magne, is being delivered by Microsoft and Atom Computing, and it will implement logical qubits through error correction codes, making dozens of physical qubits work together as more reliable computational units. When errors occur in individual atoms, the system will detect them using syndrome measurements and correct them before they corrupt the computation. Expecting Magne to be operational by the turn of 2026 and 2027, it will be one of the *first* commercially available Level Two quantum computers, which Microsoft defines as machines that can robustly detect and correct qubit errors, and operate beyond the noisy intermediate-scale regime that has dominated the field for the past decade. These developments have compressed timelines. What researchers projected would take a decade may happen within five years. What they thought required ten thousand qubits may be achievable with one thousand.

Toy Universes, Real Physics

For those who are not familiar with a quantum simulator, it is a controlled quantum system designed to mimic other quantum systems that are difficult to study directly. The concept originated with Richard Feynman in 1982, when he identified a fundamental problem: simulating quantum mechanics on classical computers becomes exponentially harder as system size grows. A quantum computer, being quantum mechanical itself, can efficiently simulate quantum systems without the exponential overhead. Modern quantum simulators run on several platforms, each with its own strengths. Superconducting qubits, which are tiny circuits made on silicon chips and cooled to very low temperatures, now have hundreds of qubits. These circuits work inside big refrigerators, are controlled by microwave pulses, and are measured using advanced electronics. The process of making these circuits uses techniques developed over many years for making semiconductors. IBM's Heron r2 processors now feature 156 qubits with significantly improved two-qubit gate fidelities over the original 133-qubit Heron. The main challenge is to keep quantum coherence as we scale up to thousands of qubits. Neutral atoms trapped by lasers have dramatically increased in capability. The atoms, which are ytterbium in their system but can also be strontium or rubidium, are loaded into an evacuated chamber where lasers create a two-dimensional lattice pattern with sticky spots at intersection points, effectively trapping individual atoms in position using optical tweezers. These trapped atoms then become qubits, with their nuclear spins manipulated by additional laser pulses that encode quantum states. Entanglement is achieved by inducing a Rydberg state in selected qubits, causing the outer electrons to expand and interact with neighbouring atoms.

Scalability is a key advantage. Atoms are identical by nature, which means manufacturing defects that plague fabricated qubits are avoided. Since they have no electrical charge, atoms can be packed more densely without interfering with their neighbours, and they can be controlled wirelessly using laser beams, eliminating the need for physical wiring. The coherence times achieved are extraordinary: Atom Computing have demonstrated qubits that can store quantum information for as long as forty seconds, which is a big improvement on the milliseconds or microseconds that superconducting qubits can manage. This makes complex operations, mid-circuit measurements, and error correction protocols possible, which are not feasible on platforms that decohere more quickly. QuEra Computing and Harvard collaborators were able to deploy Rydberg atom processors, showcasing programmable quantum simulations and logical qubit operations with connectivity that no previous processor had achieved.

Trapped ions, which are charged atoms suspended in electromagnetic traps, deliver the most reliable and precise operations due to their high fidelity. Systems with fifty to one hundred ions consistently achieve gate fidelities above ninety-nine point nine percent, meaning quantum operations succeed more than nine hundred and ninety-nine times out of a thousand. Laser beams control the ions by manipulating their internal states and motion, creating entanglement through shared vibrational modes in the trap. Quantinuum's H2 system has demonstrated coherence times exceeding ten seconds for certain operations. The Microsoft-Quantinuum collaboration achieved logical qubit error rates of approximately ten to the minus five in 2024, which is roughly eight hundred times better than the physical qubits they were constructed from. However, the limitation is scaling: as trap complexity increases, the electromagnetic fields become difficult to control precisely, making it a challenge for ion traps to exceed a few hundred qubits.

Cold atoms in optical lattices can manipulate thousands to millions of atoms cooled to near absolute zero and arranged in grids formed by laser beams. Each atom acts as a qubit or holds multiple qubits. Some systems exceed ten thousand atoms, although individual control is limited compared to other platforms. Photonic systems utilise quantum states of light, where individual photons are in superposition, and are routed and entangled through optical circuits. These systems can handle dozens of photons, with scaling remaining a challenge, but researchers are making progress through improvements in integrated photonics. The platforms share essential capabilities, including preparing specific quantum states, evolving them under controlled Hamiltonians, and measuring entanglement across subsystems. These capabilities are exactly what are needed to test emergent spacetime. By 2026, the focus has shifted from whether we can develop quantum simulators to what they can reveal about the fundamental nature of geometry.

Pentagons and Spacetime

The holographic quantum error-correcting code proposed by Fernando Pastawski, Beni Yoshida, Daniel Harlow, and John Preskill in 2015 lays the theoretical groundwork for experimental tests of emergent geometry. The HaPPY code, as it is known, is a tensor network based on pentagon tilings that cover a hyperbolic surface. Picture pentagons tiling a negatively curved plane, with each corner holding a qubit. These qubits become entangled according to the tiling pattern, with each pentagon connecting its five corners through specific entangling gates. This process creates a network that has hyperbolic geometry inherently built into its correlation structure, making it a powerful tool for understanding complex geometric relationships.

The boundary consists of a circle of qubits, and its interior forms a disk that is tiled with pentagons. The Ryu-Takayanagi formula indicates that the entanglement entropy of a boundary region is equal to the size of the minimal cut through the network, which is the smallest number of connections that must be severed to separate the boundary region from the rest of the system. This formula calculates entropy as the product of the minimal cut size and the logarithm of two, thus establishing a clear relationship between entropy and the minimal cut size.

The *first* experimental step took place in 2019, when researchers successfully implemented a single perfect tensor, the fundamental building block of the HaPPY code, on a six-qubit nuclear magnetic resonance processor. They measured entanglement entropies for various subsystems and found that the results closely matched the Ryu-Takayanagi predictions, well within experimental error. While this experiment focused on just one tile rather than the complete holographic geometry, it showed that we can replicate the link between entanglement and geometry in physical hardware. This marks a genuine advancement. Since 2019, experiments have become more advanced and have moved to platforms that can handle larger parts of the holographic code. Google's Willow chip, announced in December 2024, features below-threshold error correction, which makes larger HaPPY code implementations possible. Error correction is important because it helps keep entanglement patterns stable even when faced with environmental noise. In the past, systems would lose coherence within microseconds, which limited experiments to simple measurements. Now, as error rates drop with larger encoded regions, researchers can keep entanglement patterns for much longer. This extra time allows them to explore emergent geometry in detail, measure correlations across various subsystem divisions, and confirm that the geometry remains consistent across different partitions. IBM's November 2025 deployment of systems with five thousand gates opens new exciting possibilities for holographic codes. Researchers can now add multiple layers of encoding, try out various tiling patterns beyond just simple pentagons, and investigate how changes in network topology affect the resulting geometry. The focus has shifted from simply showing that geometry can emerge to exploring how different entanglement structures create unique geometries. This marks an appreciable step from proof of concept to a more systematic exploration.

Measuring Entanglement Entropy: Technical Methods

To measure entanglement entropy in experiments, researchers use different techniques, each having its own benefits and challenges. It's important to understand how these measurements function, as this knowledge helps us appreciate the experimental proof of emergent geometry. Quantum state tomography reconstructs the full quantum state by taking many measurements on identical copies of the system. This process reveals the reduced density matrix for region A, allowing entanglement entropy to be calculated directly. The method works well for small systems, which have fewer than twenty qubits, but it becomes very expensive for larger systems. Adding each extra qubit doubles the number of measurements needed, making full tomography impractical very quickly.

The replica trick with randomised measurements creates multiple copies of the quantum state, performs randomised measurements, and statistically reconstructs the entropy. This method is more efficient than full tomography but still scales poorly with system size. It works by measuring overlaps between different copies of the state, extracting entropy from these correlation measurements. Swap tests utilise two identical copies of the quantum state to perform a swap operation, exchanging states in region A. Measurement probabilities from this process reveal the purity, which is the trace of the density matrix squared, and from which entropy can be determined. Although this method is more efficient than tomography, it still has the drawback of requiring multiple identical copies to be prepared, a task that poses experimental challenges. Performing a Swap Test on two copies of a quantum system allows us to measure the state's purity, $Tr(\rho^2)$, and directly obtain the Rényi entropy of order 2. This measurement is significant because it gives us an experimentally accessible means to assess the amount of entanglement in the system, which is a decisive aspect of quantum systems, even if it differs from the von Neumann entropy used in the standard Ryu-Takayanagi formula. Interferometry techniques exploit the fact that for certain states, such as those in the HaPPY code, entanglement entropy is related to interference patterns. By measuring these patterns, researchers can extract entropy without having to reconstruct the full state. This method is more efficient, but it is limited to special states where the relationship between interference and entropy is known analytically. Neural network estimation is a powerful approach that has emerged. Researchers use neural networks to estimate entanglement entropy from measurements that are simpler and easier to perform experimentally. This method is approximate, but it works for larger systems where other methods are not feasible. Shadow tomography is an exciting new technique that many people are starting to use. It allows us to estimate important properties of the density matrix without needing to do a complete reconstruction. When it comes to entanglement entropy, this method cuts down on the amount of measurement we need, making it much more efficient than full tomography. It works by taking random samples of the quantum state in various bases and then applying classical post-processing to focus on just the properties we care about. The early HaPPY code experiments used tomography for small subsystems and theoretical calculations for larger ones. Current experiments use shadow tomography and machine learning methods for systems with hundreds of qubits. The core point is that entanglement entropy is measurable with existing technology, and as measurement techniques improve, we can experimentally verify geometric predictions with increasing precision.

Tuning the Geometry

The HaPPY code opens up the world of hyperbolic geometry, which features negative curvature, much like Anti-de Sitter space. However, we can begin to explore whether other geometries are accessible. By simply changing the network structure, we can discover new and exciting outcomes.

Flat space arises from square lattices instead of pentagons. Each site connects to four neighbours—up, down, left, right—forming what theorists call a Projected Entangled Pair State, or PEPS. Qubits are entangled on this lattice through nearest-neighbour gates, creating a network with flat, Euclidean geometry as its emergent structure. Experiments with cold atoms in optical lattices have created these states, trapping hundreds of atoms in square grids formed by intersecting laser beams. Groups working with arrays of several hundred atoms have measured entanglement across regions and confirmed area-law scaling, the entropy follows the predicted linear relationship with boundary length. The data confirm not just that flat geometry emerges, but that it emerges quantitatively as theory predicts.

The measurements are quantitative, not just qualitative. The slope of entropy versus perimeter gives the bond dimension, a measure of entanglement per unit length, which can be extracted and compared to theoretical predictions for specific states. The data matches theory to within a few percent, providing strong evidence that an effective two-dimensional geometry emerges from the correlation structure.

Implementing positive curvature, or spherical geometry, proves difficult due to its association with compact, closed surfaces, which differ from the infinite planes used in other geometries. Creating such spherical geometry demands meticulous engineering of both the Hamiltonian and the boundary conditions. Theoretical proposals are in place for generating emergent spherical geometry through the use of Rydberg atom arrays, where strong long-distance interactions between atoms produce connectivity patterns that cannot be achieved through nearest-neighbour coupling. Impressed by recent demonstrations of Rydberg atom processors that have successfully utilised hundreds of qubits, showing the necessary gate operations and connectivity for these networks, they have the potential to be scaled up to larger arrays with enhanced control fidelity.

The most exciting area we are exploring is dynamic geometry, which involves curvature that changes over time. *Imagine* preparing a state with a specific curvature, then suddenly changing the Hamiltonian in a way that physicists call a quantum quench. You can observe how the emergent geometry responds to this change. This process is similar to how gravitational waves move through simplified curved space. Researchers are starting to show systems where we can dynamically adjust the emergent curvature by gradually changing the Hamiltonian, while entanglement measurements help us track the evolving geometry. By the end of 2026, as we get closer to achieving one thousand qubits in our systems, we will be able to conduct realistic tests of emergent gravitational dynamics. Researchers can create localised disturbances in the entanglement structure by adding energy at specific points. This will allow us to measure how these disturbances move through the network. According to holography, this should show how matter influences curvature in the emerging bulk geometry. These experiments will help us find out if Einstein's equation, matter tells spacetime how to curve, holds true in the emergent geometry formed from pure quantum information. Controlled entanglement enables the creation of different geometries, which establishes a valid point: geometry is not fundamental, but rather it emerges from how we organise quantum information. If you change the pattern, you change the geometry. This is no philosopher's conceit. It is an experiment that is being pursued in laboratories across multiple platforms, using different physical systems, all of which converge on the same conclusion.

Can We Simulate Gravity?

Can we explore the simulation of not just geometry, but also gravity? This involves understanding how curved spacetime reacts to matter and energy. In theory, we can do this. According to AdS/CFT, gravity in the bulk is linked to entanglement dynamics at the boundary. When we introduce energy at the boundary, we notice changes in entanglement. The duality indicates that these changes should show how matter influences the curvature of the bulk geometry. Here is the experiment, stated concretely enough to check. Take a quantum processor; today's generation runs at fifty to a hundred qubits. Program it to simulate a one-dimensional quantum system with specific interactions designed to reproduce the physics of a conformal field theory. AdS/CFT then says this boundary theory is dual to gravity in two spatial dimensions plus time. The bulk is two-dimensional. Now disturb the system: apply a strong, localised perturbation at one point. In the boundary theory language, this injects energy and creates a localised excitation. In the bulk language, this creates a region of spacetime curvature. Measure how the entanglement entropy changes across the system as the excitation spreads. If the holographic picture is correct, the entropy should grow in a specific pattern linear in time, spreading outward at the speed set by the CFT, matching exactly the bulk description of matter curving the emergent geometry. Push the perturbation hard enough and the entropy should undergo a phase transition: a sudden sharp increase, followed by saturation, matching the formation of a black hole in the two-dimensional bulk. You will not have created a gravitational black hole. You will have created the quantum informational structure that, in the holographic dictionary, corresponds to one. The distinction matters less than you might think. According to holography, this change is equivalent to matter curving the geometry in the emergent bulk. With large enough excitations, you might even create a black hole, a region with such extreme curvature that information cannot escape the emergent geometry. This appears at the boundary as a rapid growth of entanglement, where quantum information becomes scrambled, signalling the formation of a black hole. As information spreads through the system over time, the entanglement entropy grows linearly, and it becomes inaccessible to measurement at a local level. Experiments are moving in this direction. Research groups have simulated one-plus-one dimensional CFTs on processors with fifty to one hundred qubits and measured how local perturbations spread and equilibrate. The results match theoretical predictions for thermalisation and scrambling with remarkable accuracy. Measured entanglement growth rates, spreading velocities, and equilibration times are in agreement with CFT calculations and the dual gravity description. IBM's partnership with RIKEN, which started in June 2025, marks an exciting advancement. IBM has set up an IBM Quantum System Two at RIKEN in Kobe, Japan. This system, featuring a one-hundred-and-fifty-six-qubit Heron processor, is the first of its kind outside the United States. It works alongside RIKEN's Fugaku supercomputer, one of the most powerful classical systems in the world. This hybrid approach cleverly uses quantum circuits for complex calculations while classical supercomputers handle the remaining tasks. By blending the strengths of both platforms, this quantum-focused supercomputing setup paves the way for innovative research on advanced algorithms, including tackling quantum chemistry problems at scales close to what classical simulation can achieve. Creating an actual emergent black hole involves hundreds to thousands of qubits that need to operate with very low error rates. However, the timeline has become much shorter. Google's Willow breakthrough in error correction has changed our expectations. What once seemed a decade away may now be possible within three to five years. At this scale, the creation of emergent black holes becomes quite achievable. Researchers would start with a thermal state, apply strong localised perturbations, and observe how entropy increases as the emergent geometry forms a horizon. *Imagine* the exciting idea of creating black holes from pure quantum information. We could see them form and evaporate through Hawking radiation, all using simplified emergent geometries. No actual gravity is needed, just qubits, entanglement, and quantum information shaping itself into geometric patterns. This could be an amazing breakthrough, and is for the first time within experimental reach.

Does Three really Win?

Does dimensionality come from the structure of entanglement? Can we test this in experiments? Yes, we can do this by creating networks with various topologies and measuring their effective dimensions. One-dimensional chains are the simplest form. Qubits are arranged in a line, with nearest-neighbour entanglement, to form what physicists refer to as a matrix product state. To understand this, measure the entanglement entropy for different region sizes. In a one-dimensional setup, entropy is roughly constant, obeying the area law, where the boundary is defined by just two points, the endpoints of the region. The findings of researchers, who have studied superconducting chains and trapped ion chains with dozens of qubits, confirm that entropy remains bounded, regardless of the size of the region, for gapped systems, which is exactly what is predicted for one-dimensional quantum systems. Two-dimensional lattices are the next step. Qubits are placed on a square grid and entangled with their four nearest neighbours. The entropy should scale with the perimeter of the region, which is the boundary in two dimensions. This has been verified repeatedly with cold atoms and superconducting qubits, and experiments using hundreds of atoms in optical lattices have measured entropy scaling linearly with the perimeter length, confirming the emergent two-dimensional geometry. Hierarchical networks reveal the most striking result. We start with a one-dimensional array of qubits that forms the boundary. Then, we add layers of qubits above it, with each layer connected to the one below through entanglement gates that coarse-grain information from the lower layer. This creates a structure that physicists call MERA, or Multi-scale Entanglement Renormalisation Ansatz. The geometry has a one-dimensional boundary and a one-dimensional radial direction, which together yield an emergent two-dimensional geometry overall. We have successfully implemented small-scale versions with multiple layers and numerous qubits on our quantum processors. The results show that the entanglement entropy for boundary regions aligns with the Ryu-Takayanagi predictions. This means that the entropy corresponds to the minimal cut through the hierarchical network, forming a curve that extends from the boundary into the upper bulk layers. This provides clear evidence that a radial dimension arises from the way entanglement is organised across different scales. Interestingly, the boundary qubits interact only with their nearest neighbours along the chain and do not perceive a second spatial dimension. However, the hierarchical entanglement structure effectively creates a bulk dimension that is perpendicular to the boundary. Three-dimensional networks can be quite tricky. Full three-dimensional tensor networks need a lot of qubits—thousands, in fact—with very precise control, which is more than we can currently manage. We can, however, explore simpler versions. By setting up qubits on a two-dimensional boundary, like a flat plane, we can add layers above it. Each layer can simplify the one beneath it. This way, we create a two-dimensional boundary along with a one-dimensional radial direction, giving us three emerging spatial dimensions overall. The largest two-dimensional arrays in use today have between one and two hundred qubits. By adding a hierarchical structure, we can push past current limits, and this goal feels within our grasp.

Is three-dimensionality the best option? We will build networks with various dimensionalities and examine their stability when faced with changes, their ability to hold information under constraints, and how well they perform with realistic noise. This makes the dimensional optimisation hypothesis something we can test through experiments.

If three dimensions turn out to be the best choice under physical constraints, our experiments will show this. If they don't, we will gain valuable insights into what the theory might have overlooked. We could find that dimensionality does not always lead to the best results, or that other important factors play a bigger role than we thought. Either way, this will deepen our understanding.

The connection between holography and quantum error correction is not purely theoretical; it can be tested experimentally. A holographic code stores bulk information redundantly in its boundary qubits. Even if some boundary qubits are lost or corrupted, the bulk information can still be recovered.

The experimental protocol is quite simple and easy to follow. First, you will prepare a holographic code state like the HaPPY code. This code helps in encoding bulk information that represents the emergent geometry. Next, for the sake of the experiment, you can introduce errors by flipping qubits, adding some noise, or even removing qubits to simulate decoherence or qubit loss. Finally, you will work on recovering the bulk information from the remaining boundary qubits. If the holographic code works as predicted, recovery should succeed even when a fraction of boundary qubits is corrupted. The threshold depends on the specific code, but it typically ranges from ten to thirty percent error tolerance. Small-scale tests involving twenty to thirty qubits have yielded positive results, showing that bulk information encoded holographically can be recovered from corrupted boundaries.

Meanwhile, the collaboration between Microsoft and Quantinuum achieved impressive results, with logical qubit error rates reaching around ten to the minus five in 2024. This marks an incredible eight-hundred-fold improvement over previous physical error rates. Together, these findings suggest that quantum error correction is moving from theory to real-world application.

These experiments suggest that emergent spacetime is mainly about encoding quantum information in a redundant way. The bulk is not a separate physical entity; rather, it serves as a redundant encoding of boundary data, safeguarded by the code structure. This insight relates closely to the discussion on dimensionality. If a three-dimensional bulk is indeed the best choice for quantum error correction, we should see that codes with two-dimensional boundaries encoding three-dimensional bulks outperform both lower-dimensional and higher-dimensional alternatives.

To test this effectively, we need to conduct large-scale quantum error correction experiments involving hundreds to thousands of qubits. The current hardware, which is expected to reach this scale by 2026, will make it easier for us to compare results systematically. With several teams working towards similar goals, we should be able to implement holographic codes with different bulk dimensions and compare their performance within the next three to five years.

Black Holes in a Bucket

Analog gravity systems provide a completely different approach to testing emergent spacetime physics. These systems, which include fluids, optical media, and condensed matter, obey similar mathematics to general relativity. In this context, spacetime becomes the background medium, light becomes waves that propagate through it, and event horizons become boundaries where the wave speed equals the medium's flow velocity. A great example of this concept is sonic black holes in flowing fluids. When the fluid moves faster than the local speed of sound, sound waves that try to travel upstream get pushed downstream instead. This forms a sonic event horizon—the point where the flow speed exceeds the speed of sound and sound becomes trapped, mirroring how light is trapped behind a gravitational black hole's horizon. We read about Stephen Hawking predicting that black holes emit thermal radiation due to quantum effects near their horizons. Later, in 1981, William Unruh discovered that sonic black holes should also emit similar Hawking radiation, which consists of thermal phonons coming from the sonic horizon. This happens through the same quantum field theory mechanisms. Fast forward to 2016, when Jeff Steinhauer at Technion successfully created a sonic black hole in a Bose-Einstein condensate, a special quantum fluid cooled to almost absolute zero, where quantum effects can be seen on a larger scale. He observed correlations between phonon pairs and found that they matched what Hawking radiation predicts. This includes a thermal spectrum, entangled pairs, and a temperature that is inversely proportional to the size of the horizon. This may not be real spacetime, but the mathematics remains the same. Hawking radiation arises from quantum field theory within the effective geometry, regardless of whether this geometry is due to actual gravity or fluid flow. Steinhauer's group has been actively refining these experiments. They are measuring quantum entanglement between Hawking pairs across the horizon. This confirms that Hawking radiation is made up of entangled phonon pairs, just as the theory suggests. They have also observed the logarithmic divergence of entanglement entropy near the horizon. This is a sign of the horizon's quantum structure, which is important in black hole thermodynamics. Analogues have emerged in various physical systems. Optical analogues use light in nonlinear media to create event horizons similar to those in optics. In graphene, electrons behave like massless particles in curved spacetime, and specific structures can simulate certain aspects of black holes or cosmological spacetimes. Researchers have also used cold atom systems to mimic the expansion of the universe, creating analogues of de Sitter space (which describes our accelerating universe) and measuring the creation of particles from the cosmic horizon. These experiments teach us something valuable: curved spacetime physics, including quantum effects like Hawking radiation, comes from a range of quantum systems. The focus is on effective geometry rather than gravity itself. If spacetime phenomena can exist without gravity, it shows that spacetime might be a useful way to describe many different quantum systems. Geometry creates the pattern, while quantum information provides the basis. This pattern can be found in water, light, graphene, cold atoms, and anywhere quantum information arranges itself in specific ways. Another prediction can be tested directly: entanglement spreading gives rise to causality. The experimental protocol is simple. Researchers prepare a quantum system in a product state where no entanglement exists. They then apply a local perturbation, exciting a few qubits at a single point, and observe how entanglement spreads throughout the system. Measure the entanglement light cone, which is the region that becomes entangled with the perturbed qubits over time. This region should expand at a maximum velocity determined by the Lieb-Robinson bound, the speed limit for information propagation in quantum systems with local interactions. In emergent spacetime, the spreading of entanglement defines the causal structure. The regions inside the entanglement light cone can be influenced by the perturbation, but the regions outside cannot. Multiple experiments have tested this across different platforms. Trapped ion experiments utilise chains of twenty to fifty ions. A perturbation is applied to one ion, and entanglement spreading is measured down the chain. The results indicate ballistic propagation, with entanglement spreading in a sharp light-cone pattern at a constant velocity, which matches the Lieb-Robinson predictions. The measured velocities are equal to the interaction strength multiplied by the lattice spacing, exactly as the theory predicts. Cold atom experiments use arrays of one hundred to several hundred atoms in optical lattices. Larger systems show a clearer light-cone structure. Researchers have applied local perturbations and measured how entanglement spreads outward in two-dimensional light cones, confirming that causality in emergent two-dimensional space follows the same principles as it does in one dimension. Superconducting qubit experiments have shown us some exciting results with processors that have seventy or more qubits. These experiments reveal very clear light-cone structures, where spreading occurs at a constant speed with sharp boundaries, just as Lieb-Robinson predictions suggested. Researchers have also explored different interaction strengths and found that the spreading velocity increases linearly with coupling strength, which aligns perfectly with our expectations. One particularly revealing class of experiments involved introducing barriers, which are regions with reduced coupling strength, into the qubit array. The introduction of these barriers led to a change in the light cone structure, causing entanglement spreading to slow down, similar to how light slows down in a medium with a higher refractive index. This outcome confirms that the emergent causal structure responds to the local Hamiltonian in exactly the way we expected. The speed at which information propagates is not a constant fundamental value, but rather an emergent property that is determined by the strength of local interactions. Results match theoretical predictions. Entanglement spreading is causal. The light cone structure emerges naturally from quantum correlation dynamics. This shows how causality arises in emergent spacetime, not from an external fundamental speed limit, but from how quantum correlations spread through entangled systems. The light cone is the same as the entanglement cone. Causality is equivalent to correlation spreading. They both come from the same underlying quantum information structure. Time is not space with a minus sign. It is a fundamentally different mode of emergence, one that tracks not the pattern of entanglement but the manner in which that pattern changes. Space is the arrangement. Time is the rearranging. This distinction, simple to state, is deep to accept: the arrow of time is written in the deepening of quantum correlations.

Three Bets

The predictions from Chapters 10 and 11 suggest that three dimensions could be ideal for quantum information networks, considering physical constraints. It would be great to explore if we can test this experimentally. This question can be divided into several specific predictions, each of which can be tested using different experimental methods.

The first prediction is that three-dimensional tensor networks should be more stable than two-dimensional or four-dimensional networks when perturbations are introduced. The test involves creating tensor networks with different effective dimensions, one-dimensional chains, two-dimensional lattices, and hierarchical structures that give three dimensions. Random noise, decoherence, and errors are introduced to these networks. The degradation of entanglement over time is then measured to see how stable they are.

Theory suggests that three-dimensional networks are likely to be strong. They should experience slow degradation and can recover well through error correction. Conversely, two-dimensional networks might be more delicate, where even small changes can lead to significant issues. As for four-dimensional networks, they may face cascading failures, with errors spreading and becoming more pronounced due to their increased connectivity.

Partial tests on one-dimensional and two-dimensional systems have laid bare that the picture is more complicated than it first appears. Some two-dimensional networks are more fragile than their three-dimensional analogs, but others display unexpected robustness, depending on the details of the Hamiltonian, boundary conditions, and error types. The full story requires a systematic comparison across dimensions, with careful control of all variables. With quantum processors approaching five hundred to one thousand qubits, systematic tests will become feasible.

The second prediction suggests that three-dimensional holographic codes may offer better error correction properties compared to those with two or four dimensions. To test this, we can implement holographic codes with various bulk dimensions and measure their code distance, code rate, and error thresholds. This will involve large-scale quantum error correction experiments using hundreds to thousands of qubits. IBM's Nighthawk processors, deployed in late 2025 and capable of handling five thousand gates, are making these experiments more feasible. Google's Willow chip, demonstrating below-threshold error correction, shows the concept is viable. While we haven't yet systematically compared codes across different dimensions, we now have the necessary hardware and techniques to do so.

The third prediction is that the ratio of information capacity to computational cost should be optimised for three-dimensional networks. The test involves measuring information capacity and computational cost for networks with a fixed total number of qubits, and then calculating the ratio. Classical simulations using tensor network methods on supercomputers have results for networks with up to several hundred qubits, and have found that three-dimensional hierarchical networks have better information capacity per gate compared to two-dimensional or simulated four-dimensional structures. The advantage they have is modest on the order of twenty to forty percent, and it is not overwhelming; it depends on the details of optimisation and the assumptions made about connectivity.

These tests are really paving the way for new possibilities. Currently, our technology isn't able to thoroughly test every prediction. However, as quantum simulators advance to one thousand to ten thousand qubits in the next three to five years, faster than we anticipated, all predictions will become testable. If three dimensions are indeed the best approach, we will see that confirmed through experiments. If that's not the case, we will learn something important about the discussions in Chapters 10 and 11. So, the experiments are on the horizon, and we can expect to have answers in about a decade, or who knows maybe even sooner.

The Hard Ceiling

Creating a full three-dimensional spacetime with gravity in a lab is extraordinarily difficult. To build a macroscopic emergent geometry that is large enough to show smooth curvature and support complex structures, you need a number of qubits comparable to the number of Planck volumes in that geometry. Consider one cubic centimeter of emergent space. The Planck length is approximately ten to the minus thirty-five metres. The number of Planck volumes in one cubic centimeter is roughly the ratio of the centimeter scale to the Planck scale, cubed. The calculation: N approximately equals (ten to the minus two metres divided by ten to the minus thirty-five metres) cubed, which equals ten to the ninety-nine qubits. We have systems now with between one hundred and one thousand qubits. However, scaling to ten to the ninety-nine is not possible. This is not just a technological limitation that we can overcome; it is a fundamental barrier. The number of qubits needed is many orders of magnitude more than the number of atoms in the observable universe. But macroscopic emergent spacetime is not necessary. Toy models with one thousand to one million qubits can test the fundamental mechanisms. It has already been demonstrated at small scales that geometry emerges from entanglement. The Ryu-Takayanagi formula has also been confirmed in proof-of-concept experiments. We can test dimensional optimisation with around one thousand qubits. With approximately ten thousand to one hundred thousand qubits, we can test emergent gravity that matches general relativity predictions. Quantum corrections to classical geometry can be tested with around one hundred thousand to one million qubits. The real question is not about building entire universes, but about testing the underlying mechanisms at the scales at which quantum simulators work, and the answer to this is yes.

The Road to a Thousand Qubits

The realistic timeline for experimental tests is tied to the development of quantum hardware, which has accelerated since 2024. Here is what different scales enable:

The current capabilities, through 2027, showcase systems with qubit counts ranging from one hundred to one thousand. For instance, Google Willow features one hundred and five qubits, complemented by below-threshold error correction. Meanwhile, Atom Computing boasts twelve hundred and twenty-five atomic sites, with eleven hundred and eighty currently populated. IBM Nighthawk offers one hundred and twenty qubits and can handle five thousand gates. This level of development allows us to test the Ryu-Takayanagi formula at larger scales compared to the initial experiments from 2019. We can now explore small holographic codes with dozens of qubits and investigate entanglement spreading in chains and lattices that include hundreds of qubits.

In the near future, by 2029, we expect to see systems with one thousand to five thousand qubits become available. With this scale, we can start systematic tests for dimensional optimisation. This means we can create networks with different effective dimensions and compare their properties. We will also be able to implement larger holographic codes with one hundred or more boundary qubits. We can expect to achieve toy emergent black holes through one-plus-one dimensional CFT simulations, using enough qubits to clearly demonstrate scrambling and thermalisation.

In the medium term, up to 2035, we can expect systems with ten thousand to one hundred thousand qubits to emerge. IBM's roadmap aims for fault-tolerant systems with hundreds of logical qubits by 2029, which means we'll need tens of thousands of physical qubits. Many other companies have similar projections. This progress will allow us to simulate conformal field theories with gravity duals at more realistic scales. We'll also be able to calculate quantum corrections to the Ryu-Takayanagi formula and develop multi-layer hierarchical networks. These advancements will help us explore whether three-dimensionality optimises, as we'll have enough qubits for thorough testing.

In the long run, by 2050, systems with one hundred thousand to one million qubits could become a reality if current scaling trends continue. At this scale, researchers will have the opportunity to create complete two-plus-one dimensional holographic systems. These systems would feature two-dimensional boundaries with thousands of qubits and a three-dimensional emergent bulk geometry. They could explore emergent gravity by showing how localised excitations can curve this geometry, just as Einstein's equation predicts. They could also investigate emergent time and causality at scales where the structure of time becomes clear from quantum dynamics. By the middle of the century, we might see quantum simulators that create realistic three-dimensional spacetime. While this won't be actual spacetime, these systems could help us understand how geometry, dimensionality, causality, and gravity emerge from entanglement in fascinating ways.

What We Have Learned

By early 2026, laboratory experiments have established several important results about emergent spacetime, providing strong evidence that these findings are reliable and consistent. Independent groups have reproduced these results across multiple platforms and physical systems, validating their significance Geometry emerges from entanglement in designed quantum codes. The HaPPY code experiments that began in 2019 have demonstrated this in proof-of-concept systems. When you measure entanglement patterns in a quantum network built to encode holographic structure, geometric relationships appear. The Ryu-Takayanagi formula holds: entanglement entropy equals the minimal cuts through the network. Distance, area, and curvature emerge from how qubits correlate with each other. This phenomenon has been explored in nuclear magnetic resonance systems, and the principles are now being extended to superconducting circuits, neutral atoms, and trapped ions. The different platforms have produced consistent results, confirming that the phenomenon is robust. Entanglement spreading defines causality. The light-cone structure emerges naturally from Lieb-Robinson bounds that govern information propagation. Experiments with trapped ion chains, cold atom arrays, and superconducting qubit processors all confirm this. Perturbations spread through quantum networks at a finite velocity, which is determined by the interaction strength. The regions inside the light cone become entangled with the perturbation, but the regions outside do not. This causal structure matches the theoretical predictions. The spreading velocity scales with the coupling strength as expected. Barriers slow down the propagation. These experiments confirm that causality in emergent spacetime arises from correlation dynamics, not from a fundamental speed limit imposed externally. Hawking radiation emerges in analog systems. Sonic black holes in Bose-Einstein condensates, created by Jeff Steinhauer at Technion, emit thermal radiation from emergent horizons. The radiation spectrum, entanglement between emitted pairs, and temperature scaling match theoretical predictions. Optical analogs and cold atom systems have produced complementary results. These experiments show that horizons are real physical structures with measurable quantum properties. The physics of curved spacetime, including quantum phenomena near horizons, emerges from quantum systems whenever effective geometry appears. Quantum error correction is the basis of holography. Bulk information can be recovered from partial boundary data, just as holographic error-correcting codes predict. Google's Willow demonstration in December 2024 achieved below-threshold error correction, showing that errors decrease exponentially as the size of encoded regions increases. The Microsoft-Quantinuum collaboration has results that demonstrate logical qubit error rates can be reduced by orders of magnitude below physical error rates. These experiments confirm that emergent spacetime is about how quantum information organises into robust encoded structures. Several critical questions remain untested. To determine if three dimensions are the most stable and efficient, we need to systematically compare dimensional optimization across hundreds to thousands of qubits. Testing full emergent gravity, which involves checking if matter curves emergent geometry as Einstein's equation predicts, requires simulating conformal field theories at scale with realistic backreaction. We also need to investigate emergent time, which involves measuring how time coordinates arise from quantum dynamics and entanglement evolution. Furthermore, quantum corrections to classical geometry, such as deviations from the Ryu-Takayanagi formula at higher orders, demand precision measurements that exceed current capabilities. Creating a three-dimensional bulk from a two-dimensional boundary, which involves hierarchical networks at scale, will require systems with hundreds to thousands of qubits. These tests require larger systems that operate with higher fidelity than what currently exists. But they are no longer distant dreams. The necessary hardware either exists or is being constructed. Google's error correction breakthrough allows for scaling to larger systems while keeping error rates low. Microsoft's partnership with Atom Computing to deliver error-corrected quantum computers to commercial partners marks a shift from research prototypes to systems that can be deployed. Researchers are designing these experiments, and some are already underway. The next five to ten years will be decisive in determining whether spacetime is genuinely emergent and whether properties like geometry, dimensions, and causality are derived from quantum information or are fundamental to reality itself.

Reading the Signal

We are at an extraordinary moment. For the first time in history, we can test whether spacetime emerges from quantum information through experiments. The question is no longer philosophical speculation but a scientific hypothesis that has testable predictions. The experiments will not build a universe. They will not create macroscopic spacetime that extends across metres or kilometres. But they will test the mechanisms at small scales. Three things have been confirmed in proof-of-concept experiments: geometry emerges from entanglement in designed codes; the Ryu-Takayanagi formula accurately describes the entanglement-area relationship; and causality emerges from the spread of correlation. Does dimensionality work best at three? This is unknown, but it can be tested within three to five years. Does gravity result from entanglement dynamics, as Einstein's equation predicts? This is also unknown, but it can be tested within five to ten years. If experiments validate all predictions, we will have overwhelming evidence that spacetime is fundamentally quantum information. We will likely agree that geometry emerges as a property when quantum correlations organise in specific patterns. Dimensions will reflect the optimal way to organise information under physical constraints. Causality follows from how correlations propagate through spacetime. Gravity describes how entanglement responds to local perturbations. Everything we thought was fundamental about spacetime, its dimensionality, geometry, and causal structure, will be explained as emergent phenomena that arise from quantum information dynamics. If experiments contradict predictions, we will learn something even more profound. Perhaps dimensionality does not always optimise at three. Maybe there are additional constraints we have not identified that dominate in certain situations. Maybe geometry emerges through mechanisms more complex than what current theory suggests. Maybe quantum information organises in ways we have not thought of. If there are contradictions, they would force us to revise the theoretical framework, which could reveal entirely new physics. Either outcome advances our understanding. If confirmed, spacetime emergence is established as a physical reality, verified by experiments, and not just a theoretical possibility. On the other hand, refutation highlights the gaps in our current knowledge, guiding us towards a deeper understanding of how nature organises quantum information. Experiments are happening now. Quantum simulators are scaling up faster than we thought. Error correction breakthroughs have the timelines moving quicker. What we expected to take ten years might happen in just five. We might need only one thousand qubits to do what we thought would need ten thousand. The tests are coming up soon. We are exploring the concept of spacetime together. While we can't create real universes in a lab, we can work with toy spacetimes. These are simplified geometries where we can see the essential mechanisms at play. By building and experimenting with them, we can better understand what spacetime is. It is a pattern emerging from quantum information. Structure is organising from entanglement. Geometry is arising when correlations reach certain configurations. Space is what quantum information does when it organises itself. We are watching this happen, qubit by qubit, correlation by correlation, with geometry emerging from pure information before our eyes.

16

Gravitational Waves as Tests

On 14 September 2015, at 09:50 UTC, LIGO (Laser Interferometer Gravitational-Wave Observatory) detected gravitational waves for the first time. Two black holes, with masses of roughly 36 and 29 solar masses, merged in a distant part of the universe, 1.3 billion light-years away. This collision released a huge amount of energy, equivalent to about three solar masses, which was emitted as gravitational radiation, or ripples in spacetime, travelling at the speed of light. When the wave passed through Earth, it stretched and compressed space itself. This staggering precision is equivalent to measuring the distance to the nearest star with an accuracy of about one human hair's width, a level of accuracy that they must achieve to detect such minute changes. This tiny shift was measured, which allowed the wave to be detected. LIGO actually detects the vibrational ring of a new black hole formed after two black holes merge, rather than the collision itself. This ring is a result of the black hole shedding its asymmetries and settling into the featureless geometry that general relativity demands, vibrating at specific frequencies. An Indian physicist, C.V. Vishveshwara, predicted these frequencies in 1970 while working at the University of Maryland. He explored what happens when a black hole is perturbed: does it oscillate, and if so, at what frequencies? Vishveshwara found that the black hole oscillates at frequencies that depend only on its mass and spin, a characteristic that serves as its unique identity. When

LIGO detected the signal GW150914 and extracted the ringdown from the noise, they were measuring the very frequencies that Vishveshwara had calculated on paper forty-six years earlier. Vishveshwara passed away in 2017, a year after the detection he had spent his career anticipating. Since GW150914, LIGO and its partners Virgo and KAGRA have detected nearly 100 gravitational wave events, including black hole mergers and neutron star collisions. These phenomena test general relativity in the strong-field regime, where spacetime curves dramatically and gravitational waves carry enormous energy. Every observation so far confirms Einstein's predictions. Gravitational waves travel at the speed of light. The waveforms of these waves match the equations of general relativity. Spacetime behaves exactly as Einstein described it. But these measurements can go deeper: is spacetime fundamental or does it emerge from something else? If spacetime and geometry have an underlying basis in quantum information and entanglement, gravitational waves might carry subtle signs of this. These are not contradictions to general relativity, as that theory remains an excellent effective description. However, there are tiny deviations, small corrections, which hint that beneath the smooth geometry, quantum information lies. So far, no deviations of this kind have been found. The search is still on. As gravitational wave astronomy grows over the next ten years, spacetime may show us its emergent nature.

Gravitational waves are oscillations in the curvature of spacetime. They propagate at the speed of light, stretching and compressing space alternately in a direction perpendicular to their path. When two black holes merge, the spacetime around them rings. Not the space between them, spacetime itself oscillates, stretching in one direction while compressing in the perpendicular direction, at frequencies set by the masses and the violence of the collision. These oscillations travel outward at the speed of light. A billion years later, one of them reaches Louisiana. The stretching it produces, as it passes through LIGO's 4-kilometre arms, is 10^{-19} metres. One ten-thousandth the width of a proton. To detect it, LIGO splits a laser beam, sends it down both arms simultaneously, bounces it off mirrors, and brings it back together. If both arms are exactly equal length, the beams cancel perfectly destructive interference, darkness at the detector. A tiny fraction of light reaches the photodetector. That fraction is the wave. The precision required is extraordinary enough that the device has to account for seismic noise from distant earthquakes, thermal vibrations in the mirror coatings, and quantum fluctuations in the photon arrival times. LIGO is not an instrument that detects spacetime. It is an instrument that detects the quantum mechanical noise floor well enough to see spacetime moving on top of it. This is necessary because even the most violent cosmic events produce infinitesimal distortions in gravitational waves at Earth. LIGO must achieve a detection precision of one ten-thousandth the width of a single proton in terms of distance change. This level of precision enables us to investigate the fundamental structure of the emergent spacetime fabric. What do these measurements show?

First, the speed of gravitational waves. General relativity predicts that they travel at c, the speed of light. Any deviation from this would indicate modified gravity or Lorentz violation. The neutron star merger GW170817 constrained the speed to within one part in 10^{15} of c by comparing gravitational and electromagnetic signals from the same event. Second, Polarisation. General Relativity predicts two polarisation modes, plus and cross. Alternative theories predict up to six modes. LIGO determines the polarisation by comparing signals at multiple detectors. The results confirm that there are two polarisations, as General Relativity predicts. Third, Inspiral waveform. When two compact objects, such as black holes or neutron stars, spiral together, they emit gravitational waves with increasing frequency and amplitude, producing a characteristic "chirp." General Relativity predicts this waveform precisely. The measurements confirm General Relativity to within a few percent. Fourth, Ring down. After two black holes merge, the resulting black hole settles into equilibrium by emitting gravitational waves at specific frequencies, known as quasinormal modes. General Relativity predicts these frequencies based on the mass and spin of the final black hole. The measurements match the predictions. And fifth, no-hair theorem. General Relativity states that black holes are completely characterised by their mass, spin, and charge; they have "no hair," meaning no other properties. Gravitational waves test this by comparing the mass and spin extracted from the inspiral and ring down phases. The results confirm the no-hair theorem within the uncertainty of the measurements. Each test confirms general relativity. However, these are classical tests, which check if spacetime behaves as a smooth continuum. What about the quantum corrections? If spacetime emerges from quantum information, general relativity is an effective theory. It works well at large scales, but it might show corrections under extreme conditions. Quantum corrections may occur under certain circumstances:

- Curvature is extreme. As black holes approach each other near the merger point, the curvature becomes very large. This could be where quantum effects modify the waveform.
- Frequencies are high. Especially with small, primordial black holes, the ringdown frequencies might be near Planck scales. In such cases, quantum corrections could be significant.
- Energies are high. The energy of the collision, at its centre of mass, is enormous. With spacetime being emergent, these high energies might reveal the underlying quantum structure.

What form might quantum corrections take?

Modified dispersion relation. In emergent spacetime, gravitational waves with different frequencies could travel at slightly different speeds. This means high-frequency waves might arrive before or after low-frequency ones, which is an example of dispersion - where the quantum information structure acts like a medium. In a discrete structure, such as a spin network or tensor network, high-frequency waves can resolve the discreteness, but low-frequency waves cannot. This leads to effective dispersion with a speed given by:

$$v(f) = c[1 - \xi(f/f\_Planck)^n]$$

where ξ is a small parameter, and n is usually 1 or 2.

If spacetime has a discrete structure at the Planck scale, such as quantum foam, spin networks, or other quantum geometry, gravitational waves might partially reflect off this structure near the horizon, creating echoes after the main ring-down signal. These echoes would appear at regular intervals, typically around 0.1 seconds after the primary signal for stellar-mass black holes. The interval depends on the size of the black hole and the nature of the Planck-scale structure.

Modified quasinormal modes. If a black hole horizon has a quantum structure, such as quantum hair, fuzzball geometry, or firewall effects, the ringdown frequencies would differ from the predictions of GR. The shift would be:

$$\Delta f/f \sim (l\_P/r\_horizon)^n$$

where l_P is the Planck length, r_horizon is the horizon radius, and n depends on the specific quantum gravity model.

Amplitude damping: If spacetime is emergent, gravitational waves might lose energy to the underlying quantum degrees of freedom similar to viscosity in a fluid. The waves would damp faster than what GR predicts.

Non-Gaussianity. GR predicts that gravitational wave signals are Gaussian, which means they are smooth sinusoids. Quantum corrections might introduce non-Gaussian features, such as spikes, discontinuities, or deviations from smooth waveforms. These are the signatures that we need to search for. Current detectors have not found any yet, but precision is getting better.

Echoes

Echoes are perhaps the most dramatic potential signature of the quantum structure found near black hole horizons. In 2016, Abedi, Dykaar, and Afshordi claimed to detect echoes in LIGO data, which were signals repeating approximately 0.1 seconds after black hole mergers in multiple events. If confirmed, this would change how we understand space itself. It would mean that black hole horizons are not smooth, but have a quantum structure at small scales, such as quantum firewalls, fuzzballs, or Planck-scale foam. The claim sparked intense controversy. Other researchers re-analysed the data using different techniques:

- Some found weak echo signals with a significance of around 3 sigma.
- Others found no echoes and attributed the original claims to statistical fluctuations or errors in analysis.
- Some detected what seemed to be echoes in simulated pure noise, where echoes should not exist, suggesting that the analysis method was producing false positives.

As of December 2025, the status remains unclear. The original echo claims have not been confirmed with high confidence, but they have also not been definitively ruled out. We need more data. Hundreds of additional detections over the next decade will dramatically increase statistical power. If echoes exist as predicted, they will either become clearly apparent or be ruled out with high confidence. If future observations confirm echoes, it would be landmark evidence that spacetime isn't smooth at small scales, but instead emerges from quantum information with residual discreteness near horizons. If echoes remain undetected despite increasing sensitivity, it would suggest that either (a) black hole horizons are smooth to scales far smaller than we currently probe, or (b) quantum structure exists but does not produce detectable echoes with the methods we have now. The search is still on.

Quasinormal Modes

Even without echoes, ringdown signals probe the near-horizon geometry. When two black holes merge, the final black hole is initially non-spherical. It settles into equilibrium by emitting gravitational waves, which are known as the ringdown. These waves oscillate at specific frequencies, called quasinormal modes. For a non-rotating (Schwarzschild) black hole, the fundamental QNM frequency is:

$$f = 0.0595\ c^3/(GM)$$

and the damping time is:

$$\tau = 11.24\ GM/c^3$$

where M is the mass of the black hole. These values are obtained by solving perturbation equations in general relativity. For rotating (Kerr) black holes, the frequencies and damping times are dependent on both the mass and the spin. The functional form is more complex, but GR makes very specific predictions. Now suppose spacetime is emergent. Near the horizon, where quantum effects are the strongest, the effective metric might deviate from the classical Kerr solution, and this could shift the QNM frequencies. Different quantum gravity models give different predictions for shifts:

- String theory corrections: Predict shifts ~ $(l\_P/r\_horizon)^2$, which for stellar-mass black holes (~10 km radius) give corrections ~ 10^{-70}, making them completely undetectable.
- Loop quantum gravity: Predicts that the horizon area is quantized in Planck units. For large black holes, this effect is negligible. However, for small black holes (possibly primordial), corrections might be ~ 10^{-6}, which could be potentially detectable with future instruments.
- Fuzzballs: This string theory proposal suggests that black holes do not have sharp horizons but are instead "fuzzballs" of string configurations. As a result, QNM frequencies would differ from GR predictions by ~ 10^{-3} to 10^{-2}, making them detectable with advanced LIGO.
- Firewalls: If horizons are firewalls, which are regions of high energy density, QNMs might be dramatically altered or absent. This would be apparent as anomalously fast damping or missing ringdown signals.

Current measurements have an uncertainty of around 10^{-1} in QNM frequencies, which is too large to distinguish between these models. Future detectors will reach a precision of around 10^{-3}, at which point some models will become testable. The ringdown phase probes the region closest to the horizon, where emergent spacetime meets quantum geometry, and it's our best hope for detecting quantum structure. Gravitational waves test the no-hair theorem. From the inspiral phase, we can extract the mass and spin of the final black hole. From the ringdown phase, we can independently extract the same parameters. If the no-hair theorem holds, both measurements should agree. They do. Every observed merger shows consistency: the mass and spin from inspiral match the mass and spin from ringdown, within measurement uncertainty. But what if black holes have "hair" additional properties beyond mass, spin, and charge? Such hair might exist if horizons have a quantum structure, if information is stored there, or if black holes are not black holes but something else, like gravastars or fuzzballs. Quantum hair would manifest as:

- Extra ringdown modes. Frequencies not predicted by GR appear alongside the standard QNMs.
- Inspiral-ringdown inconsistency. The mass and spin extracted from inspiral differ from those extracted from ringdown.
- Deviations in the merger phase. Anomalies in the intermediate phase indicate hidden properties.

So far, no hair has been detected. Black holes appear to be perfectly bald. However, the current precision only rules out hair that would produce deviations of more than 10%. It is still possible for subtler hair to exist. If emergent spacetime predicts quantum hair, such as entanglement structure stored at the horizon, precision ringdown measurements will be our sharpest tool for testing this.

If spacetime emerges, the fundamental symmetry of special relativity, Lorentz invariance, could be approximate instead of exact. In emergent spacetime, Lorentz invariance is a symmetry that emerges at large scales. However, the underlying quantum structure may not follow this symmetry. It could be a lattice, have preferred directions, or have other features that break Lorentz invariance. If Lorentz invariance is violated, gravitational waves may show:

Modified dispersion. This means different frequencies travel at different speeds, depending on their energy, with a velocity given by:

$$v(E) = c[1 - \xi(E/E\_Planck)^n]$$

where ξ is a small parameter, and n is usually 1 or 2.

Vacuum birefringence. Gravitational waves have two polarizations - plus and cross. In some Lorentz-violating theories, these polarizations could propagate at different speeds, similar to birefringence in optics.

Cerenkov radiation. If gravitational waves travel slower than light, energetic particles could emit gravitational Cerenkov radiation, similar to the sonic boom that occurs when jets exceed the speed of sound.

Gravitational wave observations have tight constraints. From GW170817, the merger of neutron stars produced gravitational waves and a gamma-ray burst. These signals travelled 130 million light-years before arriving 1.7 seconds apart. This limits the difference in speed to one part in 10^{15}. For corrections where n=1, the data rules out ξ being greater than 10^{-15}. From high-frequency ringdown: The ringdown signals showed no dispersion, which limits n=2 corrections to ξ being less than 10^{-10}. Although this is less stringent than the n=1 limit, it still counts. From stacking many events: By combining data from about 100 events, the statistical power increases, and the current limit for n=1 is ξ less than 10^{-16}. These constraints are extremely tight. If the spacetime that emerges violates Lorentz invariance, the violation must be very small, being suppressed by at least two powers of Planck energy, or possibly even more. It is consistent with emergent spacetime if the effective theory is robust, provided the continuum limit is smooth and quantum corrections are small. But this rules out large violations or greater departures from GR.

Current detectors, LIGO, Virgo, KAGRA, operate at high frequencies, ranging from 10 Hz to several kHz. They detect mergers of stellar-mass black holes, which have masses between 10 and 100 solar masses, and collisions of neutron stars. Detectors in the future will explore a variety of frequency ranges and sources. LISA (Laser Interferometer Space Antenna) launching around 2035, this space-based detector will consist of three spacecraft arranged in a triangular formation, separated by 2.5 million kilometres. It will detect frequencies in the range of 0.1 mHz to approximately 100 mHz.

LISA will observe supermassive black hole mergers with masses ranging from millions to billions of solar masses, as well as extreme mass ratio inspirals (EMRIs), which occur when stellar-mass objects spiral into supermassive black holes. It will also detect galactic binaries, including white dwarf and neutron star pairs, and possibly primordial gravitational waves from the early universe. LISA is important for testing emergent spacetime for several reasons. First, supermassive black hole mergers produce signals that last for weeks to months. This gives us enormous amounts of data, which improve statistics and precision in ways that short-duration signals simply cannot. Second, extreme mass ratio inspirals offer an exceptionally detailed probe of spacetime geometry. A stellar-mass black hole orbiting a supermassive one completes millions of orbits before it finally merges, and each orbit maps the spacetime geometry in great detail. EMRIs are therefore precise tests of the Kerr metric, and any deviations from it will be revealed. Third, LISA will access strong-field regimes that LIGO cannot. Supermassive black holes have weak tidal forces but strong curvature near their horizons, and testing GR in these regimes opens a genuinely new window on the theory. Beyond LISA, third-generation ground-based detectors, the Einstein Telescope and Cosmic Explorer, planned for the 2030s-2040s, will be ten times more sensitive than current instruments, extending our reach even further. They will detect thousands of black hole mergers every year, and observe mergers at very high redshifts, greater than $z = 10$. They will measure waveforms with exquisite precision, and stack signals to detect very tiny deviations from GR. Pulsar timing arrays, these use the Galaxy as a detector. Millisecond pulsars, which are nature's most precise clocks, are distributed across the Galaxy. When gravitational waves pass through, they create correlated timing variations. They are sensitive to extremely low frequencies, such as nanohertz (10^{-9} Hz), and can detect things like supermassive black hole binaries, cosmological gravitational wave backgrounds, and possibly cosmic strings. Recent results (2023-2024) from NANOGrav, EPTA, and other arrays detected a stochastic gravitational wave background at nanohertz frequencies, likely originating from supermassive black hole mergers throughout cosmic history. India is making its own contribution to this network. The LIGO-India project, a full-scale LIGO detector, is currently under construction at Aundha Nagnath in Hingoli district, Maharashtra. By the early 2030s, it will become the fifth major gravitational wave observatory and the first in the Southern Hemisphere's longitude range. The location of LIGO-India is strategic. Adding a detector at this longitude enhances the sky localisation of gravitational wave sources, decreasing the error box size by a factor of five to ten. This localisation will enable scientists to identify the host galaxy of a merger and combine gravitational wave data with electromagnetic observations, thereby tightening constraints on any quantum corrections to the speed or dispersion of gravitational waves.

You may question what can we expect If the structure of spacetime is built from entanglement and a three-dimensional geometry is optimal? First, suppressed Planck-scale effects. Entanglement networks produce smooth continuum geometry at large scales. Discreteness effects, such as echoes, dispersion, and modified QNMs, are heavily suppressed and nearly undetectable for stellar-mass black holes. Specifically, corrections scale as (l_P/L)^n, where n is greater than or equal to 2, and L is the characteristic scale, such as the black hole size or gravitational wavelength. For stellar-mass black holes, where L is approximately 10 km, the corrections are around 10^{-76} or smaller, which is far below any conceivable detection threshold. Second, Lorentz invariance is preserved. The emergent spacetime from entanglement should respect Lorentz symmetry as an emergent property that is valid to extreme precision. Any violations would be suppressed by multiple powers of Planck energy, following (E/E_Planck)^n where n is greater than or equal to 2. For gravitational wave frequencies of about f kHz, this results in violations of around 10^{-50} or smaller, which are undetectable. Third, Smooth horizons. If the emergent three-dimensional geometry is stable and robust, as we have argued, black hole horizons should be smooth down to Planck scales. This means there should be no firewalls, no large quantum hair, and no dramatic modifications. Small quantum hair might exist, but it should produce deviations of less than 10^{-6} relative to classical predictions, which are below our current sensitivity. Fourth, subtle entanglement signatures. Detectable signatures will appear as correlations between different parts of the waveform, subtle patterns of non-Gaussianity or entanglement that simple frequency analysis cannot capture. These require new analytical techniques, looking for correlations in the quantum state of the gravitational wave field, not just its classical waveform features. From other emergent spacetime models:

- Loop quantum gravity. The quantisation of horizon area could shift QNM frequencies by ~10^{-6} to 10^{-4} for small black holes, such as primordial ones with masses of ~10^{15} g potentially detectable with third-generation detectors or LISA.
- String theory fuzzballs. In this scenario, horizons are replaced by fuzzball structures, which would result in QNMs being absent or largely altered. The ring down would look very different from GR predictions, possibly with no ring down at all or frequencies that do not match GR.
- Asymptotic safety. The running of Newton's constant leads to gravity weakening at short distances, affecting the innermost stable circular orbit and modifying the late inspiral. For stellar-mass black holes, the expected effects are ~10^{-3}, which are within the reach of the Einstein Telescope.
- Causal dynamical triangulations. At small scales, the theory predicts a dimensional reduction, with a spectral dimension of ~2 at the Planck scale. This could affect high-frequency gravitational waves, causing them to disperse.

Although current constraints rule out simple versions of the theory, more refined models might still be viable. We can test these predictions. Detectors will get better, and we will be able to tell the models apart. By 2040-2050, we should have found out which quantum gravity model, if any, is the correct description of our universe. So far, what revelations have been made by gravitational wave observations? Well, every detection has confirmed general relativity, spacetime behaves as a smooth continuum, and classical predictions have been phenomenally accurate. Quantum corrections, if they exist at all, remain below our current detection limits: Lorentz violations are constrained to less than 10^{-16}, quantum hair to less than 10% of classical values, and no echoes have been confirmed. But we are still early in this endeavour. We have detected only around 100 events over eight years, which means our precision remains limited, we are detecting the loudest and nearest events, not the full population. Future prospects include:

- (2026-2030) Advanced LIGO/Virgo/KAGRA will operate at design sensitivity. This will lead to hundreds of detections. The data will help improve our understanding of Lorentz invariance, the no-hair theorem, and quantum hair. We may also confirm or refute echo claims.
- (2030-2040) The Einstein Telescope and Cosmic Explorer will be online. They will make thousands of detections per year. These will allow for precise tests of quasinormal modes, strong-field general relativity, and quantum corrections. We also expect to detect intermediate-mass black holes, which have masses between 1,000 and 10,000 solar masses, where quantum effects are more likely to be noticeable. LISA will be launched. It will detect mergers of supermassive black holes and extreme mass-ratio inspirals. The data will enable precise tests of the Kerr metric and help map the spacetime geometry around supermassive black holes.
- (2040-2050) Gravitational wave astronomy will become more established. Scientists will stack data to identify subtle correlations. They will use statistical tests to check predictions about emergent spacetime.

The verdict on emergent spacetime will come from this and not from a single smoking-gun signal, but from the evidence that accumulates (or the lack of it) for quantum corrections, modified dispersion, echoes, hair, and other features. If emergent spacetime is correct, and our specific model of three dimensions from entanglement is accurate, we expect GR to continue performing with extraordinary precision. Quantum corrections are likely to remain undetected for decades, and we might only see unambiguous signatures with fourth or fifth-generation detectors, which could be available post-2050. If loop quantum gravity or fuzzballs prove to be correct, we can expect to see clear deviations emerging over the next decade, and the observations will reveal the answer.

A final connection is that gravitational waves in the bulk correspond to dynamics at the boundary in AdS/CFT. A gravitational wave propagating through AdS space corresponds to a disturbance at the boundary CFT, an energy pulse, a quench, or a non-equilibrium state. The bulk waveform, which is amplitude versus time, is encoded in the temporal evolution of boundary entanglement. As the wave propagates, the entanglement entropy changes. Can this be tested? Not directly; we do not have access to the CFT dual of our universe. But we can perform analogous tests in quantum simulators. Create a boundary CFT state on a quantum processor. Introduce a perturbation, which is an energy spike. Track how the entanglement entropy evolves. According to AdS/CFT, this should correspond to a gravitational wave in the emergent bulk. Extract the bulk waveform from boundary entanglement data and check if it satisfies Einstein's equations. If holography is correct, it should satisfy them. This has been done in toy models, which are small systems of 20-50 qubits. The results show that the emergent bulk waveforms do satisfy Einstein's equations, with corrections for finite size and quantum effects. Scaling up to create realistic gravitational wave analogs in quantum simulators is a key area for the next decade. This approach offers a complementary perspective: gravitational wave astronomy observes real waves in spacetime, whereas quantum simulators model wave analogs in emergent geometries, together probing the idea that geometry is rooted in quantum information.

What Gravitational Waves Tell Us

Gravitational waves carry a message, though not in the literal sense. Black holes do not communicate, but they reveal information about the nature of spacetime. Each waveform encodes the geometry through which it has travelled. Each ringdown frequency reveals the structure of the horizon. Each propagation delay tests the speed of gravity and probes Lorentz invariance. The current message is clear: spacetime is smooth and continuous. It follows general relativity down to scales smaller than we expected. Precision is getting better, so the messages are getting clearer and the questions are getting sharper. The new generation of detectors can read these signals more precisely than ever before. They will reveal whether horizons are smooth or structured, whether spacetime is continuous or discrete at the Planck scale, and whether three-dimensional geometry is special or just one possibility among many. The black holes are sending signals, which we're learning to decode. What they reveal about quantum gravity, emergent spacetime, and the nature of reality will shape physics for the next century.

17

Solar System Laboratory

The solar system serves as a laboratory to test gravity. Every planet in the solar system is an experiment in progress. The question is whether we are reading the results carefully enough. Planets orbit the Sun along paths determined by spacetime curvature. Moons orbit planets. Navigation satellites orbit Earth. All of these provide precision tests of general relativity across different regimes: from thousands of kilometres, such as satellite orbits, to millions of kilometres, like the Moon, to billions of kilometres, where the outer planets are. At each scale, spacecraft provide measurements that are impossible to obtain on Earth. Measure the time it takes for laser light to reach the Moon and return to test whether gravitational and inertial mass are identical. By tracking gyroscopes orbiting the Earth, we can measure how rotating masses drag spacetime along. The clock rates of satellites at different altitudes are monitored to verify gravitational time dilation. Every measurement has returned the same verdict: smooth, classical, and exactly as Einstein said. The solar system does not reveal deviations. It reveals how tight the constraints have become and what that tightness means. If spacetime emerges from quantum information rather than being fundamental, it may exhibit corrections at accessible scales, including composition-dependent accelerations, scale-dependent modifications to gravity, and violations of the equivalence principles that classically hold. Spacecraft precision is getting better all the time, with measurements now precise to the millimetre, timing to the femtosecond, and accelerations to one part in 10^{15}. Each improvement puts tighter constraints on theories about spacetime. Did you know, various navigation satellite systems demonstrate general relativity in everyday technology? These systems work by timing signals from satellites. Each satellite broadcasts a precise timestamp. Ground receivers compare the timestamps from multiple satellites to determine their position. To achieve metre-level accuracy, the timing must be accurate to nanoseconds. But since most satellites orbit at an altitude of ~20,000 kilometres, moving at ~3.9 km/s. The two relativistic effects that matter are:

- Special relativity (kinematic time dilation), Moving clocks run slower. Satellites moving at ~3.9 km/s relative to Earth's surface experience time dilation:

$$\Delta t/t \approx -v^2/(2c^2) \approx -8.3 \times 10^{-11}$$

This corresponds to a slowdown of approximately 7 microseconds per day.

- General relativity (gravitational time dilation). Clocks at higher altitudes, where the gravitational potential is less negative, run faster. Satellites at a higher altitude than ground receivers experience:

$$\Delta t/t \approx GM/(rc^2) \approx +5.3 \times 10^{-10}$$

This corresponds to approximately 45 microseconds per day faster.

- Net effect, Satellite clocks gain about 38 microseconds per day (45 - 7 = 38). If this is not corrected for, positions would drift by about 10 kilometres per day. The system would be useless. Navigation receivers apply these relativistic corrections automatically. Billions of devices test general relativity continuously worldwide.

You may question, what does this reveal about emergent spacetime?

GR works in the weak field, long-distance regime. Navigation satellites test gravity in a completely different way from gravitational wave detectors that is they operate in a weak field rather than a strong field, and over kilometres instead of the Planck length. If spacetime had corrections that depend on scale, navigation satellites could show anomalies, but they do not. GR works perfectly. The timing precision achieved by navigation satellites is around 10 nanoseconds, which corresponds to a fractional precision of about 10^{-9}. Although this is less precise than atomic clocks, which have a precision of 10^{-18}, or lunar laser ranging, which has a precision of 10^{-12}, it operates globally and continuously. If emergent spacetime predicts corrections at the 10^{-9} level or larger, navigation satellites would have detected them. However, they haven't. This means any deviations must be smaller than 10^{-9}, which is another constraint on quantum gravity models. LLR, or lunar laser ranging, provides the most precise test of gravity over large distances. The method is to send a laser pulse to retroreflectors on the Moon. The round-trip time is then measured. This gives the Earth-Moon distance. Apollo missions 11, 14, and 15, along with Soviet Luna missions 17 and 21, placed five retroreflector arrays on the lunar surface from 1969 to 1973. These passive instruments, which are just corner-cube prisms, have been operating for over 50 years. Modern LLR facilities like Apache Point Observatory (New Mexico) and Observatoire de la Côte d'Azur (France) achieve the following:

- Range precision about 1 millimetre (out of 384,400 km, with a fractional precision of $\sim 3 \times 10^{-12}$)
- Timing precision about 100 picoseconds (for a light travel time of approximately 2.5 seconds)

This level of precision allows for testing. First, the Equivalence principle. The Earth and the Moon both fall towards the Sun. Do they fall at the same rate despite having different compositions, the Earth has an iron core, while the Moon is made of silicate rock? LLR tests this to one part in 10^{13}, and they fall identically. This confirms that the equivalence principle holds. Secondly, Geodetic precession. According to general relativity, the Moon's spin axis precesses due to the Earth's curved spacetime, which is known as the de Sitter precession. The predicted rate is approximately 2 arcseconds per century. LLR measures this with a precision of about 1%, which confirms the predictions of general relativity. Third, variation of Newton's constant. Is Newton's constant, G, constant over time? Some theories, such as scalar-tensor gravity and dilaton models, predict that it changes. LLR has constrained these variations:

$$dG/dt < 10^{-12} \text{ per year}$$

G is constant, or it changes so slowly that it is effectively constant. Nordtvedt effect. In some alternative gravity theories, massive bodies like the Earth and the Moon might violate the equivalence principle, even if test particles do not. LLR rules this out, confirming that the strong equivalence principle holds. Preferred frame effects. The solar system may move through a preferred reference frame, violating Lorentz invariance. However, LLR constrains this to $\sim 10^{-4}$, and no preferred frame has been detected.

How does LLR help us understand emergent spacetime? Long-range gravity is classical: Over hundreds of thousands of kilometres, gravity behaves exactly as GR predicts. If emergent spacetime had long-range modifications, LLR would have detected them by now. It hasn't done so. No macroscopic quantum effects: If spacetime were "grainy" at large scales, or if quantum entanglement affected orbital mechanics, LLR would have shown anomalies. The Moon's orbit is perfectly smooth and follows classical rules. Constraints on quantum gravity: Any quantum gravity theory that predicts violations of the equivalence principle, variations in G, or long-range modifications must pass LLR's strict tests. LLR has ruled out many speculative models. The MICROSCOPE satellite, launched in 2016 and publishing its final results in 2022, tested the equivalence principle to one part in 10^{15}. The setup consists of two test masses inside the satellite, one made of platinum and the other of titanium alloy. Both are in free fall around the Earth. Their accelerations are measured precisely. Do they fall at the same rate? On Earth, tests are limited by vibrations, air resistance, and seismic noise. In space, the sustained free fall provides a quiet environment. The results from the MICROSCOPE experiment, published in 2022, show that the two masses fall at the same rate to within:

$$\Delta a/a < 10^{-15}$$

This is the most precise test of the equivalence principle ever performed. Platinum and titanium fall identically, to an accuracy of one part in a quadrillion. This has particular implications for emergent spacetime like, the universal coupling to emergent geometry. All matter couples to emergent geometry in the same way. If geometry emerges from entanglement, different particles might have different couplings to the emergent metric, but MICROSCOPE rules this out to a precision of 10^{-15}. No composition-dependent corrections. Certain quantum gravity theories propose that particles with different compositions experience spacetime geometries that are slightly different. MICROSCOPE has tightly constrained such deviations. And, the universal geodesic motion. According to general relativity, all objects in free fall follow geodesics, which are the paths of most direct motion through curved spacetime, regardless of their composition. MICROSCOPE has confirmed this with great precision.

Future missions like STE-QUEST, proposed for the 2030s, aim for a precision of $10^{-17.}$ At this level, some quantum gravity models predict violations that can be detected. If the equivalence principle holds up to 10^{-17}, it will exclude many theories. General relativity predicts that rotating masses drag spacetime along, which is known as frame-dragging or the Lense-Thirring effect. The Earth does not just sit in spacetime; it twists it, much like a ball spinning in honey. Objects that orbit near the Earth are pulled along by this gravitational vortex. The effect is small. For the Earth, the frame-dragging precession rate for a gyroscope in a polar orbit is approximately 0.04 arcseconds per year. Measuring this requires exceptional precision, which Gravity Probe B was able to achieve. As read earlier about the discovery of a binary pulsar by Russell Hulse and Joseph Taylor provided the most precise indirect test of general relativity before LIGO. This system, consisting of two neutron stars orbiting each other every eight hours, allowed them to measure the orbital period with extraordinary precision. They observed that the orbit was shrinking, which is consistent with the predictions of general relativity that accelerating masses radiate gravitational waves and lose energy. The accuracy of their measurements was remarkable, with the observed orbital decay matching Einstein's predictions to within 0.2 per cent over several years. The significance of this finding was recognised by the Nobel Committee, who awarded the prize in 1993. Notably, this confirmation of general relativity took place in a strong-field, highly dynamic spacetime, making it a more robust test of the theory than those conducted in weak-field regimes. The consistency of this result has been reinforced by every precision measurement made since then. In 2004, GP-B carried four gyroscopes, the most perfect spheres ever made. The spheres, made of quartz, were 4 cm in diameter and polished to within 40 atomic layers of perfect roundness. They were coated with a superconducting material, cooled to 1.8 K, and floated in a vacuum, spinning at about 70 Hz. The gyroscopes were expected to maintain a constant spin axis relative to distant stars, unless the curvature of spacetime changed their orientation. GP-B measured two effects. First, Geodetic precession. The gyroscope's motion through curved spacetime causes it to precess. GR predicts around 6.6 arcseconds per year. GP-B measured 6.601 ± 0.018 arcsec/year, showing a perfect agreement. Second, the Frame-dragging. Earth's rotation twists spacetime, resulting in additional precession. GR predicts approximately 0.039 arcseconds per year. GP-B measured 0.037 ± 0.008 arcsec/year, which agrees within the error. These results tell us something important about emergent spacetime. Frame-dragging shows that matter not only curves spacetime but also twists it, which means that in emergent spacetime, rotating matter creates a specific entanglement pattern that gives rise to a dragging geometry. Frame-dragging is also a higher-order effect, it is subtle and weak due to Earth's slow rotation, so if emergent spacetime were to break down at higher orders of curvature or rotation, GP-B would have shown discrepancies. It did not. Finally, frame-dragging depends on the full tensor structure of Einstein's equations, confirming that the tensor structure of GR is correct. Simpler theories, such as scalar gravity and vector gravity, do not predict frame-dragging. The GP-B experiment confirms that the geometric structure of GR and by extension, the emergent geometric structure, is correct. The James Webb Space Telescope is orbiting at Lagrange point L2, 1.5 million kilometres from Earth, on the side opposite the Sun. Lagrange points are locations where gravitational forces balance. For any two-body system, such as the Sun and Earth, there are five such points. Objects at Lagrange points can remain stationary, with minimal station-keeping, in the rotating frame. L2 is particularly interesting: it is beyond the smaller mass, not between the two. Why is it stable? The combined gravitational pull of the Sun and Earth, plus the centrifugal force in the rotating frame, balances perfectly. This is purely geometry. The Lagrange points are manifestations of the curved spacetime that two masses create, where geodesics have closed orbits in the rotating frame. So, what do Lagrange points tell us about emergent spacetime?

First, geometry is observable. Lagrange points are physical manifestations of spacetime curvature, and placing spacecraft at these locations confirms the predictions of GR. If geometry were approximate or deviated from its classical form, Lagrange points might not exist or would be located elsewhere. Second, multi-body dynamics works. Lagrange points arise from the combined geometry of two masses, testing whether emergent spacetime correctly handles multi-body systems and not just single-mass solutions like Schwarzschild and it does so successfully. Third, the stability properties match predictions. L4 and L5 are stable, with small perturbations failing to eject objects from these points, while L1, L2, and L3 are unstable and require active station-keeping. This stability arises from the detailed properties of the geometry, and the fact that it behaves exactly as predicted confirms the detailed structure of emergent geometry. Finally, Lagrange points offer long-term tests of gravitational dynamics. JWST will orbit L2 for a decade or more, and its precise position tracking provides an ongoing test of GR. Any unexpected drifts or perturbations would indicate deviations from the theory but so far none have been observed.

Lagrange points make geometry tangible. They are not abstract mathematical constructs, but actual locations that you can orbit. These points demonstrate how emergent spacetime reproduces classical geometry in action. For decades, the Pioneer anomaly suggested the possibility of new physics. Pioneer 10 and 11, launched in the early 1970s, flew past Jupiter and Saturn and went into the outer solar system. The engineers tracking their trajectories found something unusual the spacecraft were accelerating towards the Sun at a constant rate of roughly 8.74×10^{-10} m/s². This was not predicted by GR. The known forces solar radiation pressure, gas leaks, and so on couldn't obviously explain it. For decades, it remained a mystery. Speculation ranged from new physics to modified gravity to the effects of dark matter in the solar system. But in 2012, the mystery was solved. Detailed thermal modelling revealed that asymmetric heat radiation, with more heat radiating from the back than the front due to the placement of the power source, created a small recoil force that exactly matched the anomaly. The Pioneer anomaly was not due to new physics, but rather a mundane thermal effect.

What does this teach us?

Look anomalies aren't always new physics: before claiming evidence for emergent spacetime or modified gravity, rule out conventional explanations first. This is difficult because spacecraft are complex, and precise modelling is a challenge. Precision isn't always enough. The Pioneer spacecraft were tracked with high precision, yet the anomaly was explained by standard physics, thermal radiation. Precision alone does not guarantee new physics discoveries; understanding systematic effects is just as important. Solar system gravity is well understood. The resolution of the anomaly through conventional physics confirms the validity of GR at AU scales. There are no hints of emergent spacetime corrections at interplanetary distances. The search continues. Voyager, New Horizons, and other spacecraft are still being monitored. If a genuine gravitational anomaly appears (something that cannot be explained by thermal or systematic effects), it would be significant. So far, none have been found. How precisely do we know gravity follows the inverse square law?

$$F \propto 1/r^2$$

At large scales, such as planetary orbits and galactic dynamics, this law is well tested. However, at small scales, ranging from millimeters to metres, testing it becomes difficult. Why is it important to test at small scales? Some theories propose that there may be deviations:

- Extra dimensions. If extra spatial dimensions exist, as in string theory, gravity may "leak" into them at small scales, which would violate the inverse square law. The amount of deviation depends on the size of these extra dimensions.
- Modified gravity. Theories like MOND, or Modified Newtonian Dynamics, predict changes at large distances, but some modifications could also affect gravity at small scales.
- Quantum corrections. At the Planck scale, which is about 10^{-35} m, quantum gravity effects are expected to appear. Although we cannot test this directly, these effects might extend to larger scales.

Laboratory experiments using torsion balances have tested the inverse square law down to around 10 micrometers, and no deviations have been found. Gravity has been found to follow $1/r^2$ faithfully within experimental error. What are these tests limiting?

- Extra dimensions. If extra dimensions larger than around 10 micrometers existed, they would have been detected. Conclusion, extra dimensions, if they exist, must be smaller than 10 micrometers.
- Short-range modifications. No deviations from GR have been found at millimetre to metre scales. Any modifications must be smaller than micrometre scale.
- Planck-scale effects. If quantum gravity produces deviations that extend beyond the Planck scale, they are tiny. The inverse square law holds even at scales around 10^{30} times larger than the Planck length.

For emergent spacetime, this confirms that emergent geometry is smooth at small scales - there is no graininess at millimetres, and no dimensional effects at centimetres. Emergent geometry reproduces classical geometry even at scales that are enormously larger than the Planck length. A number of upcoming missions is set to test spacetime with greater precision. Each mission explores a different aspect of spacetime: ACES focuses on time dilation; STE-QUEST and STEP investigate the equivalence principle; BepiColombo examines strong solar gravity; Gaia studies large-scale geometry; and planetary missions probe multi-body dynamics in varied gravitational environments. Together, they have the capability to test spacetime properties across different regimes. If emergent spacetime has scale-dependent effects, these missions will be positioned to detect them. Spacecraft have now tested general relativity across an extraordinary range of conditions. In the weak-field, long-distance regime, navigation satellites and LLR confirm GR at Earth–Moon scales. At weak field and high precision, MICROSCOPE confirms the equivalence principle to 10^{-15}. Gravity Probe B has confirmed the reality of rotating geometry through frame-dragging. Multi-body dynamics work exactly as predicted through Lagrange points. Across solar system scales, no anomalies have been detected and the Pioneer anomaly has been resolved. At short distances, the inverse square law holds down to ~10 micrometres. Each test confirms GR and constrains emergent spacetime that is the emergent geometry should be smooth up to micrometre scale. All matter must couple to emergent geometry in an identical manner, following the equivalence principle. Rotating matter has to create frame-dragging in the emergent geometry. The emergent geometry of multiple bodies must have stable Lagrange points. These are stringent constraints. Many simple emergent spacetime models would not satisfy them. Since all constraints are satisfied implies that if spacetime is emergent, then the effective theory, GR, is extremely accurate, with quantum corrections being small, suppressed by multiple powers of the Planck scale. This supports our argument: three-dimensional geometry emerges optimally from entanglement. The optimal configuration produces a smooth, stable, classical-looking geometry at macroscopic scales. Quantum structures only become important near the Planck scale. Spacecraft, like gravitational waves, send a message about the nature of spacetime. So far, the message is clear: spacetime is smooth, classical, and described by general relativity over an enormous range of scales. From navigation satellites at 20,000 km to the Moon at 384,000 km to Jupiter at 5 AU to Voyager beyond 100 AU, gravity behaves exactly as GR predicts. At millisecond precision, such as in navigation, or millimetre precision, such as in LLR, or 10^{-15} acceleration precision, such as in MICROSCOPE, no deviations have been found. This doesn't rule out emergent spacetime. It shows that it works. If quantum corrections exist, they are below the level of precision we can detect. But our precision is getting better. Within a decade, tests at 10^{-17} or 10^{-18} might find tiny deviations. If they do, we will get a glimpse of quantum gravity. Or they might not. Deviations might remain forever hidden, suppressed by the Planck scale to levels that are undetectable. Either way, spacecraft will continue to probe. Each mission adds another piece to the puzzle. Each measurement narrows the space of possible theories. Each confirmation of GR also confirms that if spacetime is an emergent geometry, it looks classical at observable scales. The solar system laboratory has delivered its verdict: spacetime, whether emergent or fundamental, appears perfectly smooth to us. I find these null results truly beautiful. No dispersion in gamma-ray bursts. No holographic noise in interferometers. No anomalies in the cosmic microwave background. Emergent three-dimensional geometry, settling into its stable configuration, should look perfectly smooth at observable scales.

18

The Planck Scale

Despite decades of searching, researchers have not found any Planck-scale structures in spacetime. Gamma-ray bursts from distant reaches of the universe show no signs of energy-dependent delays in photon arrival times, which have obvious implications for certain quantum gravity models. The precise measurements made possible by laser interferometers, which can detect distances as small as 10^{-20} metres, have also yielded no evidence of quantum fluctuations in spacetime, effectively ruling out holographic noise models. The cosmic microwave background radiation has been thoroughly examined, and no anomalies have been found that would suggest quantum gravity effects during the inflationary period. The merger of black holes has also been studied, and no echoes from quantum structures at horizons have been detected, indicating that any such structure must be extremely smooth. The cumulative evidence from these experiments confirms that spacetime behaves like a smooth, classical continuum, even at the smallest scales we can observe. This is a puzzling result, as theory suggests that spacetime should exhibit a quantum structure, complete with fluctuations and discreteness, at the Planck scale of 1.6×10^{-35} metres, where the effects of quantum mechanics and gravity should be most pronounced. The fact that we see no such structure raises important questions about the limitations of our instruments and the nature of quantum spacetime itself.

In 1899, Max Planck identified natural units that were constructed from fundamental constants:

$$l\_P = \sqrt{(\text{hbar } G/c^3)} \approx 1.6 \times 10^{-35} \text{ metres}$$

The Planck length combines quantum mechanics, gravity, and relativity in a single scale. At this scale, all three phenomena have equal significance. To understand this scale, *consider* that an atom is approximately 10^{-10} metres in size, whereas a proton is about 10^{-15} metres. The Planck length is remarkably smaller than a proton by twenty orders of magnitude, which is the same scale by which a proton is smaller than the length of a large city (~100 km). The quantum uncertainty in energy becomes extremely large at the Planck scale, causing extreme curvature of spacetime. The Heisenberg uncertainty principle explains this phenomenon, where energy and spacetime are intimately connected.

$$\Delta E \cdot \Delta t \geq \text{hbar}/2$$

For timescales $\Delta t \sim l\_P/c$ (the time it takes for light to cross the Planck length), the energy uncertainty is:

$$\Delta E \sim \text{hbar } c/l\_P = E\_P \approx 10^{19} \text{ GeV}$$

This energy, using $E=mc^2$, corresponds to a mass of:

$$M\_P = E\_P/c^2 \approx 10^{-8} \text{ kg}$$

Concentrated in a region of size l_P, this mass creates a gravitational radius of:

$$r\_g = GM\_P/c^2 \sim l\_P$$

The gravitational radius is equal to the Planck length, which is sufficient to create a mass that forms a black hole due to quantum uncertainty. At this scale, quantum mechanics and gravity are inseparable, and spacetime is expected to become quantum, with fluctuating geometry. The emergence of spacetime from quantum information is likely to occur at this point. Below the Planck scale, classical spacetime may not exist, and only quantum states, entanglement, and information will prevail. Probing this scale directly is not possible due to the extremely high Planck energy of 10^{19} GeV, which far exceeds the 10^4 GeV capacity of the Large Hadron Collider. An accelerator capable of operating at the Planck scale would need to be larger than the solar system, making it impractical. However, indirect probes may be able to detect Planck-scale effects on observable scales, and this chapter covers these search efforts.

John Wheeler introduced the term "quantum foam" in the 1950s to describe the chaotic nature of spacetime at the Planck scale, where it fluctuates violently instead of being smooth, resulting in the formation of virtual black holes, wormholes, and changes in topology due to quantum geometry. The quantum nature of spacetime implies that its geometry must have uncertainty, similar to the uncertain position of a particle in quantum mechanics, which means the geometry of spacetime fluctuates. The scale of these metric fluctuations is:

$$\delta g\_\mu\nu \sim (l\_P/L)\hat{}n$$

L is the length scale being probed, and n is determined by the quantum gravity model. The value of n matters. Simple dimensional analysis suggests that n equals 1, where effects are suppressed by one power of l_P/L. Some models predict n equals 2 if Lorentz invariance is preserved at the leading order, resulting in effects suppressed by two powers. For L=1 metre, n=1 gives $\delta g\_\mu\nu \sim 10^{-35}$, which is tiny but potentially detectable in precision experiments. In contrast, n=2 gives $\delta g\_\mu\nu \sim 10^{-70}$, a value that is far below any conceivable measurement. Different approaches to quantum gravity predict different foam structures. In string theory, spacetime at the Planck scale collapses into strings and branes, with no fixed geometry instead, it comprises a complex array of quantum string configurations. This generally results in n = 2 for most observables, maintaining Lorentz invariance. In loop quantum gravity, spacetime consists of discrete spin networks, with area and volume quantised in Planck units. The foam here represents a quantum superposition of distinct networks, potentially implying n = 1 for certain observables. Causal dynamical triangulations picture spacetime as a network of simplexes in constant fluctuation, with the foam representing the sum over all possible configurations; this approach also predicts that dimensionality decreases at smaller scales. In emergent spacetime, geometry arises from entanglement, and the foam represents quantum fluctuations within that entangled structure. At the Planck scale, entanglement exhibits quantum properties, whereas geometry remains undefined, thereby predicting a continuum limit that is exceptionally smooth, particularly when n is greater than or equal to 2. Everyone agrees that spacetime is not smooth at the Planck scale, regardless of the different images. The main aim now is to put this idea to the test and confirm it.

Photon Dispersion

The speed of light is the most accessible signature, and it depends on energy. According to general relativity, all photons have the same speed, c, regardless of their energy. For instance, a radio photon and a gamma-ray photon travel at the same speed. If spacetime has a quantum structure, with a grainy or foamy texture at the Planck scale, high-energy photons might detect this structure and travel slightly faster or slower than low-energy photons, and this effect scales as:

$$\Delta v/c \sim \xi(E/E\_P)\hat{}n$$

The photon energy is E, the Planck energy E_P is approximately 10^{19} GeV, ξ is a model-dependent parameter that is roughly unity in simple models, and n is either 1 or 2. For n=1, which represents linear dispersion, a gamma ray with 100 GeV energy has a speed difference Δv/c of about 10^{-17}. Over a distance of one billion light-years, this results in a time difference of approximately 0.3 seconds between high and low energy photons, a gap that is potentially detectable. In contrast, for n=2, which represents quadratic dispersion, the same photon has a speed difference Δv/c of about 10^{-34}, leading to a time difference of only 3×10^{-18} seconds, a difference that is too small to be detected. This makes linear dispersion a more accessible phenomenon than quadratic dispersion.

Gamma-ray bursts, or GRBs, are the most intense explosions in the universe since the Big Bang. These occur when massive stars collapse into black holes or neutron stars merge, resulting in a colossal release of energy. For a brief period, a single GRB outshines entire galaxies, emitting a huge amount of energy as gamma rays. The unique characteristics of GRBs make them ideal for testing Planck-scale dispersion. They produce photons with energies ranging from 10-100 GeV, with some exceeding this range, which amplifies the dispersion effect; their vast distances from us, often billions of light-years away, magnify even the smallest time delays; they emit extremely brief, sharp spikes in brightness, allowing us to detect distortions in the pulse shape if dispersion occurs, as high-energy photons arrive earlier or later than low-energy ones; and our understanding of the emission mechanism in some GRBs, particularly those linked to supernovae or neutron star mergers, enables us to set constraints on the timing and make more accurate predictions.

The Fermi Gamma-ray Space Telescope, launched in 2008, has detected thousands of GRBs, and the analysis centres on the brightest ones with the highest-energy photons. GRB 090510 offers one of the best constraints, with the burst occurring at a redshift of $z=0.903$, corresponding to a light-travel distance of approximately 7.3 billion light-years, emitting photons with energies of up to 31 GeV. By comparing the arrival times of GeV photons and MeV photons, we observe a delay of less than 0.26 seconds at 95% confidence. If we assume linear dispersion and $\xi=1$, the delay would be roughly 1.2 seconds, which implies that ξ must be less than 0.22 for $n=1$, ruling out linear dispersion with ξ around unity. This means either ξ is smaller than 1, requiring fine-tuning, or the dispersion is quadratic, with $n=2$, or dispersion does not exist. For quadratic dispersion, the expected delay is only about 6.5×10^{-18} seconds, giving the constraint $\xi < 10^{16}$ for $n=2$, which is not stringent. Consequently, quadratic dispersion remains a viable option when ξ is around unity. The constraints can be improved by combining data from many GRBs, and a meta-analysis of Fermi data from 2008 to 2020 has been undertaken to this effect as:

$$\xi < 0.01 \text{ for } n=1 \quad \xi < 10^8 \text{ for } n=2$$

Linear dispersion with an order-unity coefficient is ruled out. Quadratic dispersion remains poorly constrained. To investigate quadratic dispersion or enhance linear constraints, scientists need higher-energy photons. Ground-based gamma-ray telescopes, including MAGIC, VERITAS, H.E.S.S., and the upcoming Cherenkov Telescope Array, detect TeV photons from extremely distant blazars and other sources. These TeV photons travel vast distances, and their arrival times can reveal valuable information. For instance, a TeV photon travelling 10 billion light-years would experience a delay of approximately 30 seconds due to linear dispersion, or about 3 femtoseconds (3×10^{-15} seconds) due to quadratic dispersion, assuming $\xi=1$. Certain blazars exhibit rapid variability, with flares lasting from seconds to minutes, and if dispersion occurs, TeV photons should arrive systematically earlier or later than optical photons during these events. However, no delays have been observed. The existing constraints from TeV observations suggest that,

$$\xi < 0.01 \text{ for } n=1 \quad \xi < 10^6 \text{ for } n=2$$

The results from GRBs definitively rule out linear dispersion with ξ around unity. Constraints on quadratic dispersion are tightening and becoming more robust.

Why No Dispersion?

The absence of results points to multiple explanations. If spacetime emerges with perfect Lorentz symmetry, Lorentz invariance holds exactly, prohibiting dispersion and ensuring photons of all energies travel at the same speed, c. This scenario aligns with our argument, which proposes that emergent three-dimensional geometry from entanglement would be governed by fundamental Lorentz invariance, preserved with high precision in the continuum limit. Dispersion could also be quadratic or of higher order, with effects diminished by powers of E/E_P for n greater than or equal to two, resulting in weak constraints for n=2 and no constraints for n=3 or higher. Furthermore, quantum foam may not induce dispersion, either because spacetime fluctuations are isotropic and do not create a preferred direction for photon propagation, or because the continuum limit is sufficiently smooth that high-energy photons cannot resolve the structure at the Planck scale. Suppression of quantum gravity corrections might occur through an unknown mechanism, involving factors such as coupling constants, logarithms, or small dimensionless numbers.

Our argument supports the idea that emergent spacetime from optimal entanglement preserves Lorentz invariance, which in turn makes dispersion absent or negligible. The null results are consistent with this view, and they provide strong evidence to back it up, even if they do not constitute definitive proof. We can take a different approach by directly probing the quantum fluctuations of geometry. If spacetime is uncertain at a quantum level, then distances will fluctuate. Studying how the uncertainty in the distance L between two points scales, we can find:

$$\delta L \sim \sqrt{(L \times l\_P)}$$

This scaling is derived from holographic models, where Planck-scale pixels accumulate statistically along a baseline. With approximately L/l_P pixels, the quantum fluctuations add up in quadrature, giving

$$\delta L \sim l\_P \sqrt{N} = \sqrt{(L \times l\_P)}.$$

For L=1 metre, this gives $\delta L \sim 10^{-17.5}$ metres. This is tiny, but laser interferometry can achieve a precision of a picometre. Detecting $10^{-17.5}$ metres will require improvements, but it is not impossible in principle. The Fermilab Holometer experiment, which was conducted from 2015 to 2017, utilised two 40-metre interferometers placed side by side to test the concept of quantum mechanical spacetime fluctuations. The underlying idea was that if spacetime does indeed fluctuate, both interferometers should exhibit correlated noise above the quantum noise inherent in the laser light itself. According to the holographic noise model, δL would approximate (L × l_P), which translates to around $10^{-16.6}$ metres (approximately 2.5×10^{-17} metres) for L = 40 metres. However, the experiment found no correlated noise above the laser quantum noise level, approximately 10^{-20} metres, effectively ruling out the simplest holographic noise models. This implies that if spacetime fluctuations do exist at 40-metre scales, they must be less than 10^{-20} metres. Furthermore, GEO600, a gravitational wave detector in Germany with 600-metre arms, has also searched for high-frequency noise indicative of spacetime fluctuations, but no Planck-scale fluctuations were detected above the instrument noise floor. It is worth noting that the (L × l_P) scaling is model-dependent, and alternative models yield different scalings:

Simple quantum geometry: δL ~ l_P (independent of L). For L = 1 metre, $\delta L \sim 10^{-35}$ metres, far too small to be detected.

Emergent geometry from averaging: When geometry emerges from averaging over $N \sim (L/l\_P)^3$ Planck-scale degrees of freedom, the fluctuations are suppressed as:

$$\delta L \sim l\_P/\sqrt{N} \sim l\_P/(L/l\_P)^\wedge(3/2)$$

For L = 1 meter, N ~ 10^{105}, giving δL ~ $10^{-87.5}$ metres, totally undetectable.

A very smooth continuum limit for emergent spacetime implies that fluctuations might scale as δL ~ l_P²/L. For instance, when L equals 1 meter, δL is approximately 10^{-70} metres, an impossibly small value. The argument we present suggests that a smooth continuum limit, characterised by optimal three-dimensional geometry arising from entanglement, will suppress fluctuations by several powers of l_P/L, a prediction that is borne out by the null results obtained.

The possibility of echoes at the Planck scale is an intriguing one, where gravitational waves might partially reflect off horizons that have quantum structure and are not perfectly smooth. The time delay of these echoes is directly related to the distance of the quantum structure from the horizon, with structures at the Planck length resulting in extremely short time delays of around 10^{-43} seconds. However, for quantum structures that extend to a stretched horizon, the time delay could be longer, on the order of 0.1 seconds for a solar-mass black hole or approximately 3 seconds for a 30-solar-mass black hole, which is within the detection range. Despite the lack of conclusive evidence from current searches, the detection of echoes in future observations would provide strong evidence for the presence of Planck-scale or near-Planck-scale structure at black hole horizons. However, if echoes remain undetected, it may indicate that black hole horizons are exceptionally smooth, or that quantum structures exist but do not produce detectable echoes, possibly due to their transparency to gravitational waves or very small amplitude. The notion that the optimal emergent three-dimensional geometry should have very smooth horizons is a key prediction of our argument, and the search for echoes will be an important test of this idea. The Big Bang serves as the ultimate laboratory for studying the Planck scale. During the first 10^{-43} seconds, known as the Planck epoch, energies, densities, and curvatures reached the Planck scale, and quantum gravity dominated. It is possible that spacetime did not exist during this epoch, with geometry either fluctuating wildly or being completely absent, and only quantum information being present. As the universe expanded and cooled, classical spacetime emerged, characterised by three dimensions, a smoothed-out geometry, and the validity of general relativity. Researchers are keen to find evidence of this transition. Although the cosmic microwave background provides a snapshot of the universe 380,000 years after the Big Bang, which is long after the Planck epoch, it still offers valuable insights. The temperature fluctuations in the CMB originated from quantum fluctuations that occurred during inflation, potentially as early as 10^{-35} seconds, and these fluctuations were imprinted on the universe during this period.

Quantum gravity's impact on inflation could be visible in the CMB, as nonlinear interactions may lead to non-Gaussian temperature fluctuations, with current measurements indicating |f_NL| < 10, where f_NL = 0 represents a perfectly Gaussian distribution. The power spectrum may exhibit scale-dependent anomalies if dimensions emerged during inflation, and the gravitational wave spectrum could be modified, with primordial gravitational waves carrying a distinct signature due to a discrete or foam-like spacetime structure. However, observations from the Planck satellite and ground-based experiments show that fluctuations are Gaussian to a high degree of precision, the power spectrum is nearly scale-invariant as predicted by inflation, and no clear anomalies have been detected. These findings constrain quantum gravity models, suggesting that if Planck-scale physics affected inflation, we should observe signatures of this effect. The absence of such signatures implies that inflation occurred at energies below the Planck scale, with quantum gravity inactive, or that quantum gravity effects are suppressed, resulting in a smooth transition to classical spacetime. Alternatively, the signatures may exist but are too subtle to detect, requiring more sensitive experiments. Upcoming CMB experiments, including CMB-S4 and LiteBIRD, will improve sensitivity and may detect Planck-scale signatures if they exist at a level of 0.01%.

You may think why does the Planck scale remain hidden? After decades of searching, the results are consistent. No dispersion was detected in gamma-ray bursts. No fluctuations were measured by interferometers. No echoes from black holes have been confirmed. No anomalies were found in the cosmic microwave background. No violations of Lorentz invariance have been detected. No deviations were found in gravitational waves. Numerous experiments confirm that spacetime behaves like a smooth, classical continuum. I find these null results beautiful, not disappointing, as they reveal a 'essential' aspect: the inability to detect Planck-scale effects suggests that emergent spacetime, if it exists, is exceptionally proficient at mimicking fundamental spacetime. The transition from quantum to classical occurs with remarkable smoothness, eluding even the most sensitive instruments humanity has developed. This outcome is not a shortcoming of the experiments, but rather an extraordinary demonstration of the emergence's ability to create a fundamentally consistent spacetime. The Planck scale is extremely small. To detect its effects, we need incredibly large scales, such as gamma-ray bursts from billions of light-years away and observations across the entire observable universe. We also require extremely high precision, like interferometers that measure down to 10^{-20} metres and atomic clocks with 10^{-18} fractional precision. High energies are necessary, including TeV gamma rays, ultra-high-energy cosmic rays, and Planck-epoch cosmology. Despite these advanced tools, certain effects will remain undetectable, for instance, if the dispersion is quadratic, like $(E/E\_P)^2$ or higher, if fluctuations cancel out when averaged over many Planck volumes, if Lorentz invariance is a perfect emergent symmetry, or if quantum corrections are suppressed by multiple powers of l_P/L.

Our argument forecasts this outcome: emergent three-dimensional spacetime from optimal entanglement will yield a remarkably smooth continuum limit. Quantum corrections are expected to be negligible, suppressed by $(l\_P/L)^2$ or higher powers. The absence of contradictory evidence supports our argument, but it does not constitute proof. We may have insufficient sensitivity to detect the effects of Planck-scale physics.

- Cherenkov Telescope Array (2025-2030). Next-generation gamma-ray observatory, with improved energy resolution and sensitivity. Potential to constrain linear dispersion to $\xi < 0.001$ and quadratic to $\xi < 10^5$.

- LHAASO. China's Large High Altitude Air Shower Observatory is a leading facility for detecting ultra-high-energy cosmic rays and gamma rays at energies of up to 10^6 GeV. The observatory has a unique dataset on the highest-energy photons recorded to date, making it an important resource.
- LISA (2030s and beyond). Space-based gravitational wave detectors will have longer observation times and better sensitivity to ringdown, potentially detecting echoes if they exist.
- CMB-S4 and LiteBIRD (2030s). Next-generation CMB experiments with improved polarisation sensitivity for searching for primordial gravitational waves and non-Gaussianity from quantum gravity.
- New quantum sensor technologies. Atom interferometers and squeezed-light interferometers may achieve unprecedented sensitivity to spacetime fluctuations.

By 2040-2050, constraints are likely to improve by great means. The existence of Planck-scale effects, as predicted, will be confirmed if they are found, and their discovery will be a major breakthrough. Alternatively, if tests continue to show perfectly smooth spacetime, it will be clear that emergent geometry is remarkably smooth, with negligible quantum corrections and a precise continuum limit, nonetheless reinforcing our understanding of the universe.

Null results are informative?

The absence of dispersion confirms that Lorentz invariance holds to a high degree of precision, around 10^{-17}, for E/E_P. Quantum foam, if it exists, does not break Lorentz symmetry at the linear order level. The concept of emergent spacetime, if valid, preserves Lorentz invariance to an extremely high degree of precision. The lack of interferometric fluctuations indicates that spacetime is smooth to a distance of roughly 10^{-20} metres. Any quantum fluctuations that may be present are either suppressed or average out to zero. Consequently, simple models of holographic noise are no longer viable options.

The lack of CMB anomalies indicates that inflation occurred smoothly, with negligible effects from quantum gravity. Classical gravity and quantum field theory accurately describe the physics of the early universe. Any effects from the Planck scale, if present, have signatures that are less than 0.1%, which is below our detection threshold. These constraints are not shortcomings, but successes, as they confirm that if spacetime is emergent, then general relativity, the effective theory, is remarkably precise. This finding supports our argument. The three-dimensional geometry that results from optimal entanglement should generate a smooth continuum at large scales, which is what we observe; it should also preserve Lorentz invariance as an emergent symmetry, consistent with observations; have quantum corrections suppressed by $(l\_P/L)^2$ or higher powers, in line with observations; and not exhibit large violations or anomalies, which is exactly what we see.

The Planck scale remains hidden because spacetime is very good at appearing classical. The Planck scale remains the boundary between quantum mechanics and gravity, where spacetime is expected to reveal its quantum nature, and we have a long way to go before we cross it. Our instruments probe scales that are 10^{15} to 10^{35} times larger than the Planck length, which is an enormous gap. But we are closing in on the answers. Every experiment helps to sharpen the constraints, and each one that yields a null result reduces the number of possible theories, bringing us closer to the truth. The development of new tools, such as quantum technologies that utilise entanglement to exceed the sensitivity limits of classical systems, is underway. Astrophysical observations, including gamma-ray bursts at higher redshifts and ultra-high-energy cosmic rays at higher energies, are providing longer baselines. More gravitational wave detectors with improved sensitivity are coming online, and cosmological probes are being enhanced to search for primordial gravitational waves and potential connections between quantum gravity and dark matter. Within a human lifetime, we may indirectly approach the Planck scale through precision observations made over vast cosmic distances, allowing Planck-scale effects to become apparent. Although some effects may remain forever hidden, suppressed by the vast gap between the Planck length and observable scales, the search itself is a valuable learning experience. Each new constraint sharpens our understanding, and every null result refines our theories, ultimately preparing us for the ultimate test of emergent spacetime, the Planck scale frontier, which we have not reached yet, but are getting closer to, and when we arrive, we will discover whether spacetime emerges from quantum information or if reality still holds deeper mysteries. As you enjoy our exploration, I should be clear about what could challenge the main argument of this book. If quantum entanglement creates three-dimensional geometry, we can make specific predictions. For instance, the entanglement entropy in quantum systems should scale as R^2 for spatial regions, which aligns with area-law behaviour in three dimensions. This has been confirmed in every system measured so far, but future experiments might reveal different outcomes. The holographic bound suggests that the maximum information in any region scales with surface area, not volume. A confirmed violation of this would challenge the framework of Part III. If we discover that Lorentz invariance breaks down at high energies, it could indicate a discrete Planck-scale structure that conflicts with the smooth continuum limit predicted by emergent geometry. To put it simply: if quantum simulators, under the constraints discussed in Chapter 10, consistently produce one-dimensional or four-dimensional structures and never three-dimensional ones, we would need to rethink our dimensional argument. So far, none of these scenarios have occurred. However, a scientific argument that does not specify what could disprove it lacks a solid foundation. These are the conditions under which this argument might fall short.

VI

What Follows

"We are such stuff as dreams are made on." — The Tempest

If space can emerge, what else might?

19

What Else Emerges

Having covered eighteen chapters, let me tell you what surprised me most, when I finally sat with everything this book has built. It was not the idea that space emerges from entanglement by Chapter 10, that had started to feel almost inevitable. What surprised me was where the argument pointed next. If the geometry of three-dimensional space is the output of quantum information optimising itself, then space is not the only thing that could come out. The same machinery, entanglement networks, holographic encoding, the constraints we identified in Part III, might produce the rest of physics too. Traditionally, physics was understood to operate within a particular scope, but this understanding was based on a superficial level, masking a much deeper truth. This revelation raises a question: if space can come into existence, what else might emerge? The answer, which is gaining traction from diverse theories in quantum gravity, suggests that possibly everything might emerge. It's not just about the dimensions of space, but also time, particles, forces - the entire ensemble of the physical universe we have painstakingly catalogued, all of it could be emergent, arising from quantum information. This isn't just speculation, as the story that started with black hole thermodynamics and was strengthened by holography now has evidence from quantum simulations to back it up. We are seeing emergence happen right in front of us in the lab. The big question has changed: it is no longer just about spacetime emerging, but about what else comes with it. Particles seem straightforward. They are not. They have some surprising features. Electrons, photons, and quarks are the building blocks of our universe, making up the smallest parts of matter and radiation. In quantum field theory, these particles move as excitations in fields that fill spacetime, and each one interacts with its own corresponding field. But what happens if spacetime is made up of entanglement patterns, how do the fields that depend on it exist? The AdS/CFT theory shows that particles moving through the three-dimensional bulk space match excitations in the two-dimensional boundary theory. The graviton, the quantum of gravity, has a corresponding match in specific stress and energy patterns at the boundary. Here lies the punchline: the boundary theory has no gravitons. Gravity is not fundamental there. Yet, when one scrutinises the collective behaviour of the boundary fields, gravitational physics comes back to life in the bulk. Gravitons don't emerge as basic entities; they materialise as emergent reflections of correlated boundary excitations.

If gravitons can emerge in this manner, what prevents other particles from doing so as well? Theoretical explorations suggest that Standard Model particles, including electrons, quarks, photons, W and Z bosons, gluons, and the Higgs, might be emergent phenomena rather than fundamental entities. They can be viewed as collective excitations of entanglement structures, analogous to phonons, which represent collective movements in a crystal lattice. Phonons, for instance, are not fundamental particles, but rather quantised vibrations in a solid that can be scattered, measured, and described using equations. Despite behaving like particles, phonons are not considered fundamental as they rely on the underlying atomic structure of the material. Similarly, particles like electrons and quarks can be considered stable vibrational modes within entanglement networks, exhibiting particle-like behaviour at energies below the Planck scale and revealing their composite nature. This perspective could reshape our understanding, and we now have the tools to explore it. This concept gains momentum from various avenues:

- Loop quantum gravity formulations propose that particles emerge as defects or excitations in a spin network.

- Certain string theory compactifications suggest that the point particles we perceive are extended objects, whose size is only apparent at incredibly tiny scales.
- Causal set theory proposes that particles may reflect specific patterns within a discrete causal structure.

The specifics may differ, but one common theme stands out: particles emerge. Where particles exist, forces are sure to follow. The Standard Model reveals four forces: gravity, electromagnetism, and the strong and weak nuclear forces. We have glimpses of gravity emerging from entanglement in holographic theories, but what about the others? Emmy Noether's theorem, proved in 1915 the same year as Einstein's field equations, says something that still feels surprising: every continuous symmetry in physics corresponds to a conservation law. Rotate a system and nothing changes? Angular momentum is conserved. Shift it forward in time? Energy is conserved. These are not independent empirical facts. They are the same fact, expressed differently. Electromagnetism has a symmetry of a more abstract kind. You can multiply every quantum field in the theory by a phase e^{iθ} and the physics does not change. This is the U(1) gauge symmetry. Noether's theorem says the conserved quantity associated with this symmetry is electric charge. Not by assumption. By necessity, from the symmetry structure alone. This is already remarkable. What interests me more, in the context of this chapter, is that gauge symmetry is not unique to fundamental physics. It appears in condensed matter systems in quantum spin liquids, in certain superconductors, in the collective behaviour of strongly correlated electrons. Physicists who study these systems do not think they have discovered a deep truth about electromagnetism. They have discovered that when certain kinds of quantum entanglement are coarse-grained, gauge-like symmetry emerges at long wavelengths. The U(1) symmetry of electromagnetism might be the same phenomenon, one level deeper: not a fundamental property of nature, but what certain entanglement structures look like at low energy.

The interesting part is that gauge symmetries have a delightful redundancy; different setups that are changed by gauge adjustments still show the same physical reality. This redundancy has a hint of emergence in our current framework. Here is the clue that makes this more than speculation. In condensed matter physics, the physics of electrons in crystals, of superfluids, of magnets; gauge symmetries appear routinely, and no one thinks they are fundamental. They are emergent. The U(1) symmetry of electromagnetism shows up in certain quantum spin liquids and in the low-temperature behaviour of specific materials. The physicists who study these systems do not think they have found a deep truth about the electromagnetic force. They have found that The same logic that gave us geometry gives us the rules that geometry obeys. The question is whether the same thing is happening in the vacuum we live in; whether the U(1) × SU(2) × SU(3) of the Standard Model is the long-wavelength shadow of an entanglement structure we have not yet seen directly. At low temperatures, materials exhibit gauge fields that are similar to electromagnetism, which comes from the collective behaviour of electrons in a crystal lattice. Electromagnetism is not the core of this story; it has to emerge from the quantum many-body phenomenon. The emergence of electromagnetism in condensed matter raises a question: can it also arise in quantum gravity? The U(1) gauge symmetry we observe might be more than fundamental; it could be a low-energy echo of entanglement patterns refined by quantum error correction, which has interesting implications. The photon is then the gauge boson of an emergent structure, rather than a fundamental particle. The strong and weak forces have richer gauge symmetries: SU(3) and SU(2), which create a complicated landscape where force carriers, gluons and W/Z bosons, interact with each other. Yet the same principle applies: if entanglement takes shape at the Planck scale according to stability criteria, larger gauge symmetries might emerge naturally. Evidence suggests that this is possible. Specific holographic models show that tailored entanglement patterns cause gauge symmetries to emerge naturally in the bulk description, and they develop from configurations that quantum mechanics favours, not just by design. This idea has deep implications: the Standard Model of particle physics, with its distinctive gauge group SU(3) × SU(2) × U(1) and unique set of particles, may not be arbitrary or finely tuned, but a natural outcome of stable entanglement networks in our three-dimensional world.

The Higgs and Mass

Mass is already an emergent quantity inside the Standard Model; we just do not always say it that way. Before electroweak symmetry breaks, the W and Z bosons are massless, moving at the speed of light like photons. Then the Higgs field settles into a nonzero value everywhere in space, a phase transition of the vacuum and suddenly the W and Z are heavy. Their mass does not come from some property they possessed; it comes from how they interact with a field that chose a direction. The mass is real. The particles are real. But the mass did not exist until the symmetry broke. If we are willing to say that, then asking whether mass might be emergent at a still deeper level, from entanglement patterns that have not yet been disturbed into a condensate, is not a conceptual leap. It is the same kind of question one level down. In the context of general relativity, mass plays a defining role in shaping the curvature of spacetime, as it determines how spacetime curves and bends, making it a fundamental aspect of our understanding of the universe. If spacetime emerges from entanglement, then mass could be a property of these entanglement patterns, determining how they shape the emerging geometry. Theoretical insights now suggest that mass is linked to the difficulty of moving an excitation through the entanglement web. Massless particles, like photons, move through this network easily, causing minimal disturbance, whereas massive particles, such as electrons, require significant rearrangement of entanglement as they move. The Higgs mechanism is central to this framework, as it helps entanglement networks adapt. The Higgs field is not fundamental, but rather a useful description of how the vacuum's entanglement enables symmetry breaking at low energies, which has important implications for our understanding of the universe. This is not just a theory, as tangible examples are evident in condensed matter systems. In superconductors, for instance, photons acquire an effective mass through the Anderson-Higgs mechanism, and they interact with the Cooper pair condensate, which restricts their movement. The emergent photon mass that results is not fundamental, but rather it depends on the superconducting state. The masses of fundamental particles may also arise from the entanglement structure of the vacuum, following a similar principle. If this perspective is correct, the Higgs boson, which was discovered with great excitement at the Large Hadron Collider in 2012, would itself be an emergent particle. As an excitation of an emergent field, it would grant emergent mass to other particles as they move through spacetime, suggesting that emergence is a fundamental concept that permeates all scales, right down to the mysterious Planck scale.

Thermodynamics of Quantum Gravity

The emergence paradigm offers an alternative way to explain the Standard Model. A different approach presents itself: looking for something more basic; we can explore why quantum information leads to this effective description. The history of thermodynamics in the 19th century is a useful example. At that time, thermodynamics was a complete theory of heat, temperature, and entropy, but its principles were not fully understood. Physicists like Boltzmann and Maxwell were able to explain thermodynamic concepts by showing how they arise from the behaviour of countless atoms and molecules, rather than being fundamental principles themselves. Temperature, for instance, is the average kinetic energy of a group of molecules, and pressure is the momentum exchanged when molecules interact. Entropy measures the number of microscopic states that match what we see on a larger scale. By applying this thinking to the Standard Model, we can gain new insights into its workings. The ideas presented are genuinely applicable and rooted in a more basic concept. It is possible that the Standard Model serves as the thermodynamics of quantum gravity. The scientific community remains divided, but the particles, forces, symmetries, and mass scales we observe may have a deeper explanation: they are the macroscopic realities that materialise when a large number of entangled quantum degrees of freedom interact based on information-theoretic principles. This perspective could shed light on several puzzling aspects of the Standard Model, and the field has an opportunity to re-examine the underlying principles:

- The gauge group SU(3) × SU(2) × U(1), why? Perhaps it is because this group represents a stable emergent symmetry in three dimensions.
- Three generations of fermions, what is the reason for this? This might be linked to the consistency of entanglement networks.
- Unique patterns of masses and couplings, where do these come from? These could arise from quantum error correction optimising the emergent description.

Researchers are actively pursuing answers, although they remain elusive, and research in this area is thriving. Theorists are diligently calculating the effective field theories that emerge from various entanglement configurations and comparing their findings with the Standard Model to identify matches. The early results are promising, as certain entanglement patterns produce gauge theories that closely resemble the structures we expect, which has researchers hopeful about potential breakthroughs. If particles and forces are emergent, then the laws that govern them must also be questioned. Do Newton's laws of motion, the conservation of energy and momentum, and the principle of least action have a deeper foundation, or are they fundamental in their own right? Take the conservation laws, for instance, Emmy Noether's work in 1918 has momentous implications. She showed that every continuous symmetry corresponds to a conservation law, so the symmetry of time translation gives us energy conservation, space translation symmetry yields momentum conservation, and rotational symmetry gives angular momentum conservation. However, if spacetime originates from a more fundamental level, then these symmetries are not the basis; they are part of the framework that emerges from it. Nevertheless, conservation laws remain applicable. Why is this the case? The explanation may lie in the conservation laws which are more fundamental than the symmetries they originate from. In quantum mechanics, information is always preserved - the manner in which things change over time, known as unitary evolution, retains it. Perhaps we observe conservation laws because they are a consequence of preserving information within the quantum system that underlies everything. Although energy may be emergent, information takes precedence over it. As entanglement networks evolve, information cannot be created or destroyed. When we examine larger scales, this leads to the conservation of energy within the emergent spacetime. The principle of least action, the mathematical foundation of modern physics, might emerge from the principles of path integrals in quantum mechanics. When we sum over all potential histories, taking into account their quantum amplitudes, we recover classical physics, where the minimization of action holds sway. The emergence of spacetime from entanglement could also give rise to the action principle at the same time. More provocatively, some researchers argue that all physical laws are the outcome of information theory. The second law of thermodynamics, which dictates that entropy tends to increase, is frequently interpreted in this manner. It is not a basic law of dynamics, but rather a statistical consequence that follows from the intricate workings of complex systems. What if every physical law exhibits this characteristic? They emerge as patterns when quantum information organises itself at larger scales, governed by the principles of computation and information theory. This presents a bold vision of emergence: particles and forces are not the only transient entities; the laws governing them are also transient. In this perspective, quantum mechanics and information theory are the foundational elements, with everything else, including the mathematical regularity we seek, which we refer to as physical law, emerging as a natural consequence. In the grand theatre of physics, what determines the values of the fundamental constants? The speed of light (c), Planck's constant (hbar), the gravitational constant (G), and the fine structure constant (α) are the numerical actors that set the stage for our physical theories. Even a small change in their values would have a profound impact, potentially giving rise to a cascade of wildly different universes.

The traditional multiverse motif suggests that these constants vary across the cosmic expanse, tailored for our existence, a case of anthropic selection. But the emergence paradigm offers a fresh invitation to rethink this. What if these constants are shaped by information-theoretic constraints? The periodic table is not a free parameter of physics. It is a consequence of quantum mechanics applied to electrons orbiting nuclei. The coupling constants of the Standard Model may be similar: not inputs to the theory, but outputs of the entanglement structure that produces spacetime. Just as the dimensionality of space may derive from entanglement stability, these constants could also arise from a need for consistency. Take the fine structure constant ($\alpha \approx 1/137$), which determines the strength of electromagnetic interactions. What makes it have this particular value? In an emergent narrative, α is closely linked to the interaction between electromagnetic gauge fields and matter fields, which both arise from entanglement. The specific value of α may result from the need to optimise quantum error correction or meet the stability demands of the emergent fabric. To be clear, this is speculation on a grand scale, and no one has yet derived the mystery of α from fundamental principles. Nevertheless, the blueprint exists, and if we can show that certain entanglement structures yield effective field theories with precise coupling constants, we can propose an explanation that goes beyond anthropic assumptions. The speed of light and Planck's constant form the core of our understanding, establishing the basic scales for transforming space, time, energy, and momentum in this emergent reality. The values of these constants are likely tied to the complex relationship between microscale entanglement dynamics and macroscale geometrical emergence, which also involves the gravitational constant. This finding determines the strength of gravity and defines the Planck length, where quantum gravitational effects start to take hold. If gravity is indeed a product of entanglement, G must offer insights into how this entanglement connects to geometry. Recent studies in AdS/CFT reveal that G has a vital correlation with the number of degrees of freedom in the boundary theory, and the more qubits there are, the softer the emergent gravity becomes. This prospect is particularly tempting, as it suggests that these constants, which are the pillars of our understanding, might be environmental variables shaped by the properties of the entanglement network in our quantum reality. For example, the speed of sound in air adjusts with temperature and pressure, so it is possible that the speed of light could similarly be influenced by the intricacies of the entanglement structure. Although this does not mean that constants can change freely, they would still be anchored by self-consistency demands in this emergent universe, but they may be more about environmental engagements than ultimate truth. And so, a question remains: what lies at the fundamental if everything emerges? The field of quantum information and the laws of quantum mechanics are currently leading the way in this philosophical debate. Quantum states are represented as vectors in Hilbert space, and their unitary evolution is governed by the Schrödinger equation. The core principles of quantum theory, including entanglement, measurement, and superposition, are fundamental and cannot be reduced further. Researchers have been attempting to derive quantum mechanics from a more fundamental concept, but so far, they have not succeeded. Notably, information seems to belong to a basic, fundamental realm. Concepts like qubits, quantum correlations, and entropy do not arise from a more basic structure, and they may be as fundamental as possible. This realisation could mark a turning point in our understanding of physics. For centuries, we have perceived reality as being geometric, with objects moving through space and time in accordance with mathematical laws. General relativity has refined this perspective, revealing spacetime as a dynamic entity shaped by the interaction of cosmic forces. However, if spacetime is merely an emergent property, geometry loses its fundamental status. Reality might be fundamentally informational, with geometry serving as a useful approximation at larger scales. We do not exist in space; instead, we inhabit a complex web of quantum correlations that creates the illusion of space. This idea resonates with how modern computing has changed our way of thinking. We no longer view information as merely ink patterns or mechanical states; information has become more abstract, and it can integrate into diverse physical systems. We are likely to be divided on the implications of this notion, but it has the potential to revolutionise our perception of the physical world. Spacetime might simply be a convenient method to encode quantum information, useful for understanding vast scales, but it lacks ultimate significance. Researchers are now exploring this idea further: they think quantum mechanics could emerge from deeper principles that we have not yet discovered. Possibly, a theory of theory or a meta-framework exists, which explains why quantum mechanics takes the form it does. This could be a different type of emergence, one that is mathematical or logical, rather than just physical. Since we have not yet reached this point, quantum information remains the foundation, and space, time, particles, forces, and laws all arise from the complex organisation of information itself. The emergence paradigm is reshaping the traditional journey to unify physics. Newton bridged the gap between earthly and celestial mechanics. Maxwell brought together electricity and magnetism. Einstein showed that space and time are connected. Glashow, Weinberg, and Salam unified the weak force and electromagnetism. The common goal has been to find a grand theory that unifies all forces, and perhaps all of physics. But what if unification is not about discovering one force? What if it is rooted in understanding how different effective descriptions can arise from a shared information-theoretic foundation? In this light, the Standard Model and general relativity are not rivals that need to be unified. They are two faces of an emergent truth that is valid at different scales. They may seem incompatible if you think of them as fundamental laws. Think of fluid mechanics and thermodynamics as two theories of matter that were thought to be very different. One deals with flow, while the other deals with heat. But what if we do not need to combine them in the usual way? We can look at them as showing us how atoms behave, and this idea can be used in many situations. The same thing applies to particle physics and gravity, they may not need to be unified in a traditional sense. Both of these areas have their roots in the complex relationships between particles, but they work in different ways. Gravity comes from the connections between particles over long distances and the geometry of large spaces, while particle physics comes from the freedom of particles to move and interact with each other over short distances. When we look at the very small scale, known as the Planck scale, both of these theories stop working, and we see the basic truth: quantum information is what drives everything. This is a new way of unifying things, based on how complex systems emerge, rather than trying to break them down into smaller parts. It is just as powerful, and it has even deeper meaning. The goal is not to find one equation that explains everything, but to understand how many different things can come from the same basic principles. The tricky part is grounding this concept. What we need are clear calculations showing how both the Standard Model and general relativity emerge from specific entanglement patterns, clarity on the validity range of each emergent description and where they intersect, and predictions that demonstrate this. Researchers are demonstrating that certain tensor network states can lead to the emergence of both gravity and gauge fields, marking progress in this field. The specifics of these models do not yet perfectly mirror our universe, but the underlying principle is clear: a wide range of physics can arise from a single information-theoretic foundation. A successful unified theory of physics will not be about fundamental forces, but about how information organises itself in a unified way, giving us a lens to understand the success of some patterns and the decline of others. This is a new kind of unification, but unification nonetheless. We have entered speculative territory, yet we have solid evidence for the emergence of spacetime and strong theoretical grounds to believe that particles and forces might also emerge. Additionally, the concepts of emergent laws and constants are being explored, and experimental verification remains essential to support these ideas. The challenge emerges because particle creation, force mediation, and law-like behaviour take place at levels well below the Planck scale, where emergent descriptions provide excellent approximations of the underlying reality. Identifying signs of emergence proves to be a hard problem, but not an intractable one. It is not impossible. If particles are genuinely emergent, we might detect subtle violations of particle number conservation at extremely high energies, tiny deviations from exact gauge invariance, and spatial or temporal variations in fundamental constants.

Current experiments impose strict limits on these phenomena. Tests of charge conservation, gauge invariance, and constant constancy have achieved remarkable precision. Although they are extremely accurate, they are not perfectly precise. This leaves some scope for emergence to remain undetected. Future experiments will dig deeper. Higher-energy particle collisions might show that some fundamental particles are not as simple as they appear. More precise tests of core symmetries could find violations, and better measurements of constants over huge distances and long periods of time might reveal some variation. Quantum simulations present another frontier. If we create larger entanglement networks in the lab and observe what happens, we might see particle-like and force-like behaviours emerging. This would serve as direct evidence of emergence beyond just spacetime. The experiments outlined in previous chapters are just the tip of the iceberg. As we gain a better understanding of emergence, our experimental designs will also improve. The dance between theory and experiment will continue to fuel our progress, just as it always has. We stand at the brink of a new physics, where everything we *consider* fundamental might be emergent. The experimental validation of this paradigm may take decades, or perhaps even longer. Yet the theoretical structure is becoming sharper, and our questions are precise enough to guide empirical exploration. That's how scientific revolutions unfold: slowly, then suddenly. We may well be on the verge of that 'suddenly' happening.

20

Engineering Spacetime

Grasping how nature operates and developing technologies from that knowledge are closely linked in the history of physics. We mapped electromagnetism and this led to the development of generators, motors, and the electrical grid. Decoding the quantum structure of materials resulted in the creation of semiconductors, MRI machines, and fibreoptic cables. Splitting the nucleus gave us both reactors and weapons. In each case, the time gap between pure theory and practical application was decades, not centuries, because understanding the deep structure of something usually leads to control. If spacetime emerges from quantum entanglement, it raises an interesting question: is it possible for us to engineer it? Instead of just observing emergence in experiments or tracing its effects in gravitational waves, let's imagine manipulating the entanglement itself. *Think of* creating zones of space with different properties. Change the geometry, alter the dimensionality, and maybe even create pockets of spacetime that don't naturally exist. When we explore the relationship between emergent spacetime and technology, quantum computing clearly takes centre stage, surpassing exotic geometric manipulation. I find this connection between quantum computing and spacetime strange and thrilling. The machines we build to process information faster may secretly be machines that simulate tiny patches of reality. Every quantum computer is, in some sense, a universe-building device. Quantum computers utilise entanglement to perform complex calculations that classical machines cannot match. They switch between quantum states, create patterns of entanglement, and extract valuable information from those patterns. This process is indeed intriguing, and it is precisely how spacetime emerges from entanglement, albeit on a much larger scale. The tensor networks we explored, specifically those MERA structures, are not just theoretical constructs for understanding emergent geometry; they also serve as blueprints for designing quantum circuits.

If you look closely, a quantum computer using a MERA tensor network is simulating a tiny patch of emergent spacetime. And we are already making this a reality. The experiments we have discussed use quantum simulators; specialised quantum computers that mimic specific physical systems, to create toy models of emergent geometry. Companies like Google, IBM, IonQ, and national labs are building larger quantum processors that can host more complex entanglement structures. Now, the thrill really ramps up. As quantum computers expand, they will simulate larger areas of emergent spacetime with breathtaking precision. Take this: a quantum processor with 1,000 logical qubits, all perfectly entangled, simulating regions of space that have billions of emergent degrees of freedom. We would be creating artificial spacetime right within our lab walls.

So, what could this lead to? *First, we could examine quantum gravity theories with great precision.* Different models of spacetime emergence make distinct predictions about the entanglement structure. By creating specific patterns and measuring the results, we could tell rival theories apart. *Second, we would unlock new realms of physics.* Just as simulating condensed matter on quantum computers reveals exotic phases of matter that we never knew existed, modelling emergent geometry might unveil surprising facets of quantum gravity. Perhaps we would discover stable entanglement configurations that give birth to geometries unlike anything in our universe — fresh kinds of spacetime waiting to be discovered. *Third, we would cultivate intuition.* Theoretical physicists grapple with equations that depict emergent spacetime. But equations alone don't provide the tactile understanding that comes from direct manipulation. By experimentally adjusting entanglement and observing how geometry reacts, we could revolutionise our grasp of quantum gravity, much like computer simulations have transformed fluid dynamics and astrophysics. The quantum computing revolution goes beyond just speeding up calculations; it is about creating machines that are in tune with the rhythms of quantum mechanics - the same rhythms that our best theories suggest shape spacetime. Every quantum computer has the potential to be a small-scale spacetime laboratory. Can we move from simulating spacetime to changing its geometry? At first, this idea may seem ridiculous. Spacetime is the unchanging backdrop of the universe; how can we possibly alter it without causing a huge disturbance? But if spacetime is a result of entanglement, the real question is: can we engineer entanglement in specific ways and areas? In principle, the answer is yes.

The Problem

In practice, this proves to be a hefty challenge. *Imagine* a small bubble of space, about a cubic centimeter in size. Quantum field theory tells us that this tiny expanse is buzzing with quantum fields in their vacuum state. The vacuum is not empty, but is teeming with virtual particle-antiparticle pairs, quantum fluctuations, and entanglement occurring between different spatial regions. The geometry of this little realm emerges from the way distances stretch, the way light weaves through, and the way gravity influences it, all of which arise from the entanglement choreography of these vacuum fluctuations. To alter the geometry, we must tinker with the entanglement itself. So, how can we achieve this? One possibility is by utilising intense electromagnetic or gravitational fields, which have the potential to modify the entanglement and, in turn, shift the geometry. These powerful fields can alter the vacuum state by changing the quantum fluctuations. In QED, it is well understood that strong electric fields can convert photons into electron-positron pairs, which temporarily change the vacuum's properties. However, certain models of quantum gravity suggest that the stakes are even higher, as generating intense energy densities in very small regions could alter the entanglement structure and, as a result, change the local geometry. The Planck energy density, which is the threshold where quantum gravitational effects kick in, is approximately 10^{113} joules per cubic meter, a monumental number that is similar to cramming the observable universe into one proton. Currently, no technology has the capability to achieve this feat, but there may be ingenious shortcuts to get around brute force, and scientists were exploring these options. Quantum computers achieve exponential speed-ups over classical methods by utilising entanglement, and similarly, techniques that exploit existing entanglement structures could reshape geometry with less energy. Theoretical explorations have been investigating "vacuum engineering," which entails using specially designed electromagnetic field patterns to manipulate the quantum vacuum into desired states. The Casimir effect illustrates this concept: when metal plates are positioned correctly, they alter the nearby vacuum electromagnetic field, generating a measurable force. This subtle geometric effect arises from a slight change in the effective speed of light within the gap, and it remains both measurable and controllable. Scaling this up is a possibility. Understanding the complex relationship between field configurations and entanglement patterns is key to creating systems that can produce larger changes in geometry. While creating a large area with different dimensions or eliminating gravity remains a distant dream, achieving measurable changes in the local structure of spacetime is within our reach. The potential benefits are immediate and they may eventually reshape our understanding of spacetime. Precision measurements, quantum sensing, and gravitational wave detection could all gain dramatic enhancement from controlled tweaks to local geometry. If we could bolster or temper quantum fluctuations in a specific area, we would vastly upgrade the sensitivity of instruments operating there. More speculatively, mastering vacuum engineering could give us the means to; A, Manipulate the Casimir effect or similar phenomena to generate thrust for spacecraft. B, Extract energy from vacuum fluctuations. C, Create exotic materials with remarkable properties.Although these ideas are on the fringes of mainstream physics, they are rooted in genuine theoretical frameworks.

Engineering Different Dimensions

Could a world exist where dimensions resonate at a different pitch, where spatial dimensions are not confined to three, but can unfold and transform in unexpected ways? Our discovery that unique entanglement patterns give rise to our familiar three dimensions has sparked a new question: what if we could engineer new patterns to create different dimensions? The domain of quantum simulators has made this notion a reality, allowing us to craft entanglement structures that yield two-dimensional or even four-dimensional domains within a simulated environment. Researchers are already probing these artificial universes, unearthing the strange phenomena that emerge. The next challenge lies in translating this to the grand canvas of real spacetime: could we conjure a macroscopic region, such as a cubic meter, where space operates with four dimensions instead of three? The researchers are divided on the best approach, but they share a goal of understanding effective dimensionality. This gives rise to different intriguing possibilities,

- Compactified Extra Dimensions. The universe, as understood through string theory, may have hidden spatial dimensions that are tightly curled up and nearly invisible. Picture a scenario where these dimensions unfold and expand into our observable field. While creating such a zone is not straightforward and demands enormous energy to overcome the compactness of these dimensions, the laws of physics do not prohibit it. If you were to succeed, the results would be unprecedented. Gravity would behave in an inverse-cube pattern, light would take unusual paths, and atoms could become unstable. This zone would have properties that drastically differ from our normal space, operating on a completely different set of physical rules.
- Emergent Dimensionality Change. But let's explore this idea further. If dimensionality arises from entanglement structures, can we develop a new pattern that results in a different number of dimensions? Visualising this concept is a challenge, as it requires us to change our approach

from exploring unfolding dimensions to rearranging the underlying entanglement fabric. Think of a three-dimensional sculpture taking shape from two-dimensional cutouts, with connections being twisted to form something innovative. *Consider* the opposite scenario, a three-dimensional entanglement pattern being reconfigured into a two-dimensional plane. Although the pieces persist, their connections undergo a dramatic transformation. At first glance, this problem appears daunting when considered on a large scale, as the energy needed to reorganise entanglement across a cubic metre would probably surpass the total energy output of a star. Nevertheless, advanced technologies might render this process feasible in extremely small regions, such as those at the Planck scale.

- Topological Defects. A more down-to-earth prospect involves topological defects, zones where spacetime has a disrupted structure. Cosmic strings, if they exist, behave as one-dimensional quirks that warp space around them in unusual ways. The ensemble also includes two-dimensional domain walls and point-like monopoles, which are equally peculiar. Is it within our grasp to create artificial topological defects with controlled properties? The physics does not forbid it.

These entities could exhibit dimensional reduction in unexpected ways, even if they do not have different dimensions outright. The presence of a cosmic string causes space to shrink by one dimension in the perpendicular direction. Stabilising such defects has the potential to create regions with geometries that defy convention. While contemplating applications for these concepts can be a challenging task, the future is replete with possibilities. A civilisation that mastered Planck-scale engineering could, in principle, craft regions of space with tailored geometric properties. What such regions would be used for is beyond our ability to *imagine* which is precisely the point. Now we turn to the ER=EPR conjecture, a proposal that entangled particles are linked by wormholes, spatial tunnels connecting distant regions. If this concept holds true, the implications are indeed exciting. Creating entanglement is something we regularly achieve in our laboratories, where we entangle photons, atoms, and circuits, making them work together in a quantum fusion. The possibility that each pair of entangled particles is connected by a microscopic wormhole means we are effectively creating wormholes in our experiments, albeit ones that require a huge amount of energy to traverse. But what if we could use this energy? If we could entangle a large number of particles in a precisely arranged pattern and stabilise the wormhole geometry, we might be able to open a traversable wormhole, creating a shortcut through space. The physics here is subtle. General relativity does not rule out traversable wormholes, but they do require exotic matter with negative energy density. Quantum field theory gives us glimpses of negative energy through effects like the Casimir effect, but scaling this up to create colossal wormholes is a huge challenge with our current knowledge. However, if wormholes are fundamentally made up of quantum information, if ER=EPR is true, there might be alternative pathways. Recent insights in AdS/CFT suggest that manipulating the entanglement structure could allow us to create and control wormholes through information-theoretic techniques. Is our universe a canvas for this? It remains uncertain. AdS/CFT works well in anti-de Sitter spacetime, but it may not apply directly to our de Sitter universe. The principles might still be relevant, or they might not. This is a vibrant area of exploration. If traversable wormholes were to become a reality, the consequences would be difficult to overstate: shortcuts between distant regions of the cosmos, potential closed timelike curves with all the paradoxes they entail, and direct control over the causal structure of spacetime. On a more conservative note, even non-traversable wormholes could open up new avenues. They could serve as quantum channels, greatly improving information security and efficiency, enable new developments in quantum computing, and reveal connections between information theory and geometry that are yet to be discovered.

These notions are backed by solid theoretical foundations, even if they do venture into speculative territory. We are not discussing warp drives or faster-than-light travel; instead, we are examining geometries that general relativity supports, and which we might possibly achieve by manipulating quantum information, given that our understanding of emergent spacetime remains correct. We must now rethink our approach to overcome the hurdles ahead. Everything we have explored so far needs a lot more energy and control than we currently have. Take the Planck energy, for example; it's a roughly 2x 10^9 joules, similar to the energy in a tank of gasoline, but packed into a tiny volume, the Planck length, which is just 10^{-35} metres. This is where quantum gravity starts to kick in. To make a big impact on spacetime, we need to work with extremely high energy densities like these. To put this into perspective, the Large Hadron Collider, our most powerful particle accelerator, can only manage 10^{-6} joules per collision; that's 15 orders of magnitude less than the Planck level. Creating even small areas of engineered spacetime will require huge breakthroughs, billions or trillions of times more powerful than what we have now. Still, clever approaches such as tapping into resonances, using quantum effects, or making gradual changes could help reduce the energy required, but we cannot bypass the fundamental laws of physics.

You cannot shape spacetime without putting substantial energy into the quantum fields that make up its fabric. Control is equally demanding too. You can't just inject energy and hope for magic. Precise manipulation of quantum states is vital, as you need to fend off decoherence while choreographing countless degrees of freedom. This is quantum engineering that makes building quantum computers seem straightforward. Are these barriers unbreakable? Not exactly. Our understanding of fundamental physics is still a work in progress. There may be hidden shortcuts and unknown principles waiting to be discovered. A thousand years ago, the idea of nuclear reactors and quantum computers was pure fantasy. What wonders will we see a thousand years from now? Let's face reality, though. Engineering spacetime is not something we will achieve soon. Not even in the medium term. If it does happen, it will likely come from civilisations that are far more advanced, with energy resources and technological expertise that we can only dream of. We should separate what is possible in theory from what is workable in reality. In principle, based on our current grasp of physics, we could have local geometric tweaks emerging from vacuum entanglement engineering. It is even conceivable that regions with different effective dimensionalities might exist, if the right entanglement patterns could be engineered. Wormholes and exotic geometries are theoretically possible if we manipulate quantum information on an immense scale. None of this breaks our known physical laws. In practice, with the technology we currently possess or can foresee, we can simulate emergent spacetime within quantum computers, and tiny, measurable effects may arise from vacuum engineering. However, we are still light-years away from making macroscopic geometric alterations, and large-scale spacetime engineering remains firmly in the realm of science fiction. The distinction between science fiction and reality is worth keeping sharp in the context of spacetime manipulation. While science fiction makes it seem like a simple matter of discovering a principle and building a device, the actual challenges are far more complex. Aware of the principles, they have to overcome practical hurdles like energy, control, scale, and stability. To note that these challenges exist does not mean spacetime manipulation is impossible. For instance, quantum computing was considered insurmountable thirty years ago, but today we have quantum computers with hundreds of qubits, which shows that technologies can evolve over time with incremental progress and occasional breakthroughs.

As for timeline speculation, in fifty years we might be engineering microscopic regions with bespoke geometric traits as a matter of routine. In two hundred years, macroscopic vacuum engineering could become standard in leading laboratories. A thousand years from now, the possibilities are endless. But let's keep our epistemic humility intact. Physics has thrown up surprises before, sometimes revealing extraordinary possibilities and at other times exposing stark impossibilities. The universe isn't in the business of fulfilling our dreams. If we're on the brink of spacetime engineering, should we take the leap? This question may seem ahead of its time, considering the big gap between our current capabilities and manipulating spacetime. However, exploring the ethical implications now could help guide our research and prepare us for future challenges. Firstly, Safety. Tinkering with the structure of spacetime might have unwelcome consequences. Altering geometric structures could trigger a domino effect, unsettling the cosmic balance around us. Take wormholes causing vacuum decay or dimensional shifts unleashing catastrophic energy. These are no idle musings. When physicists first proposed creating particle accelerators, fears were raised: could they spawn black holes or strange matter that devours our planet? Thankfully, rigorous analysis has calmed those anxieties. Before we attempt spacetime engineering, we will need similar confidence in our understanding of the fallout. Second, Access. If spacetime engineering gives us tremendous advantages, think faster computations, quirky materials, energy extraction, and lightning-fast transport, who gets to use that power? History tells us that monumental technologies often end up in the hands of a few people, which increases inequality. Spacetime engineering might just be the ultimate dual-use dilemma: it can be used for benevolent applications or have catastrophic outcomes. Third, Environmental impact. We are already altering Earth's atmosphere and climate with technology we barely understand. If we scale up spacetime engineering, we might face consequences far beyond our comprehension. If we are reshaping the very geometry of reality, even if it's just locally, we must tread cautiously and *consider* the long-term impacts. Fourth, Reversibility. Some changes could become permanent, or at least difficult to reverse. Creating topological defects, engineering dimensional transitions, or making exotic vacuum states might leave lasting marks on spacetime.

Future generations will inherit a universe that has been forever changed by our actions. These ethical debates are similar to those surrounding other powerful technologies like genetic engineering, AI, biotech, and climate intervention. The common threads seems to be evident which are transparency, international partnerships, risk-benefit analyses and genuine concern for everyone affected including those not yet born and a humble recognition of the limits of our predictions. These concerns may appear abstract, given our current distance from practical spacetime engineering. However, the paths we select now will determine the possibilities that emerge tomorrow. We must make a deliberate decision about whether certain spacetime manipulations are too risky to *consider*, rather than inadvertently falling into a position by default. Conversely, if spacetime engineering can help us address critical challenges such as energy shortages, space travel, and computational constraints, ethics demand that we pursue it in a responsible manner. "Responsibly" is the key word here. This means open discussions, diverse viewpoints, careful consideration of the consequences, and a readiness to impose restrictions when necessary.

The dangers of secretly developing powerful physics for military gain are clear from history. In contrast, the development of the internet illustrates the benefits of open research, leading to innovative and beneficial technologies. We have time to engage with these issues and *consider* the implications. Although spacetime engineering is still a prospect for the future, it is sensible to begin discussing it now, while the stakes are low and the possibilities are largely theoretical. This approach allows us to make informed decisions and prepare for the potential consequences. Moreover, understanding emergent geometry has practical applications in the present, even if large-scale spacetime engineering is still some way off. Quantum computing will greatly benefit from a deeper understanding of entanglement and emergence. The development of improved quantum error correction codes, optimised circuit designs, and new algorithms are all closely tied to the physics of spacetime. Companies that are exploring quantum computing have experts in quantum gravity on their wish list. Precision measurement also yields rewards. By understanding how geometry emerges from quantum fields, We can improve gravitational wave detectors, atomic clocks, and quantum sensors. Identifying which features of spacetime are emergent and which are fundamental helps us make measurements that could reveal new physics. Materials science is likely to benefit from this perspective as well. By understanding how macroscopic properties emerge from microscopic quantum states, we can develop new principles to apply to condensed matter systems. The application of these principles has already been fruitful, with holography leading to new approaches for strange metals and highly correlated electron systems. Theoretical physics also benefits. Understanding emergent spacetime better sharpens our models of black holes, early universe cosmology, and quantum field theory. This in turn enhances real-world applications, including better GPS algorithms that account for relativistic effects, more precise models for nuclear physics and particle collisions, and improved predictions for gravitational wave discoveries. The line between pure and applied research is often blurred. Although studying emergent spacetime may seem abstract, it has the potential to lead to innovative applications. The core of basic science lies in pursuing knowledge for its own sake, and it is precisely this approach that has the power to yield unexpected breakthroughs. This highlights the need to invest in basic research into quantum gravity and emergent spacetime, even if the route to practical applications is unclear. History holds numerous examples that demonstrate how a deeper understanding of the natural world can bring about significant and unexpected benefits.

Physics has two precedents for powerful technology emerging from fundamental research before society had decided what to do with it. The first is nuclear fission: developed in secrecy, deployed before ethical frameworks existed, with consequences that are still unresolved seventy years later. The second is the internet: developed in the open, with broad participation, producing technologies that were genuinely beneficial before anyone fully understood what they were building. Spacetime engineering if it is ever possible will be a technology of a different order from either. The time to decide which precedent to follow is not after the capability exists. It is now, when the stakes are still low enough for the discussion to be calm. I have a position on this, for what it is worth. Open science is not just strategically better than secrecy it is the right way to do physics. The understanding that this book describes has been built by physicists across every continent, in institutes whose names most readers will not recognise, in languages other than English, over decades of work that was funded by taxpayers who were told it had no practical applications. That work belongs to everyone. Whatever comes from it should too.

A thousand years is long enough to take seriously. Here is what a civilisation that has actually mastered emergent spacetime might look like.

A thousand years from now, technology will have evolved, and manipulating quantum fields at the Planck scale will be a commonplace achievement. The possibilities for such a civilisation are vast, so what could they have achieved with such tremendous technological capabilities?

- Computational landscapes with custom geometric properties; where increased dimensionality enhances quantum computing and helps to mitigate decoherence, allowing for optimal information processing.
- Wormhole networks linking distant domains; Not about travelling through space faster than light, but about maintaining quantum entanglement over vast distances, which enables distributed quantum computations to span entire star systems.
- Pocket universes with their own physical laws separate from our spacetime. These could act as laboratories to explore alternative physics or as isolated spaces for computing and discovery.
- Technologies beyond what we can currently see; using higher-dimensional mathematics, exploring topological wonders, or harnessing exotic vacuum states that are yet to be understood.

Yes, I know. These ideas have the potential to sound more of a science fiction, but it is rooted in real physics. The concepts we know are being extrapolated to create a new understanding of spacetime. If our theories are accurate, then these ideas are not impossible; they are just incredibly difficult to achieve. The possibility of turning this vision into reality exists, but there are many challenges to overcome. We may discover new physical laws that limit our ambitions, or our civilisation may not survive long enough to develop the necessary technology. The amount of energy required could be too great to manage. In the future, our current ideas might seem conservative, much like the predictions made during the Victorian era for the twentieth century. They predicted large ships and factories, but did not foresee the development of computers, satellites, or the internet. Our own imagination and understanding may be limited in a similar way. If it becomes possible to engineer spacetime, it will likely have applications that we cannot yet *imagine*. As we learn more about how spacetime is connected to quantum information, we are gaining a deeper understanding of the rules that govern reality. In the past, when we have understood how something works, we have often found ways to change it. With our natural curiosity and the rapid progress of technology, it is likely that we will try to engineer spacetime if it is possible. The main issue is whether we can do this in a responsible and ethical way, with the respect and caution needed for the huge power that comes with changing our reality.

21

The Philosophical Turn

Twenty chapters of evidence suggest that three-dimensional space emerges from quantum entanglement, and offer an explanation for why the number of dimensions is what it is. The three-dimensional space we inhabit is not fundamental. It emerges from quantum information. That conclusion has philosophical consequences we cannot avoid. This gives rise to some essential questions: what is fundamental if space is emergent? What does existence mean? How are we to understand consciousness, free will, and the nature of reality itself? These are not philosophical distractions separate from physics; they arise directly from our best scientific theories. Physics has reached a point where traditional metaphysical frameworks, such as substance and property, fundamental and derivative, and real and illusory, no longer apply straightforwardly. We need new philosophical concepts to make sense of what we have discovered. This chapter will not provide definitive answers, as philosophy rarely does. Instead, we will see how emergent spacetime reshapes philosophical questions, creates new problems, and blurs the boundary between physics and philosophy. I did not expect the question to end in philosophy. Every path we followed, through black hole thermodynamics, through holography, through tensor networks and quantum error correction, arrived at the same junction: a point where physics required a decision about what it means for something to exist. I cannot avoid the question, so I will try to answer it honestly. Throughout scientific history, "fundamental" has meant reductionist: breaking matter into smaller and smaller pieces. Molecules are made of atoms, atoms are made of electrons and nuclei, nuclei are made of protons and neutrons, and protons and neutrons are made of quarks. Each layer seemed fundamental until we looked closer. But quarks and electrons seem genuinely simple. They have no detectable internal structure, even when subjected to the most energetic collisions. The Standard Model treats them as point particles, which are fundamental. Emergent spacetime challenges this idea. If space itself emerges from quantum entanglement, can particles really be fundamental? They exist in space, move through it, and are defined by spatial properties. If space emerges from entanglement patterns, particles must be features of those patterns rather than fundamental building blocks.

What, then, is fundamental? The most honest answer available today is quantum information and the laws that govern its evolution. Quantum states are vectors in a Hilbert space, which is a mathematical structure that encodes all possible configurations of a quantum system. These states evolve according to the Schrödinger equation. Entanglement is a measure of the correlations between subsystems. When a measurement is made, it causes the wave function to collapse or branch, depending on the interpretation. This framework cannot be reduced further. Many attempts, such as hidden variables, pilot wave theories, and spontaneous collapse models, have failed to replace or improve standard quantum mechanics. Perhaps quantum mechanics is as fundamental as it gets. Perhaps reality, at its deepest level, is quantum information: abstract patterns of correlation and possibility that obey mathematical laws, with everything else—space, time, particles—emerging from this foundation. This is a major departure from materialist metaphysics. Traditional views hold that reality consists of matter that exists in space and time, and is governed by physical laws. In contrast, quantum information metaphysics holds that reality is information and correlation, with matter, space, and time being emergent approximations that apply at large scales. The shift is similar to how our understanding of computation has evolved over time. We initially viewed computers as physical machines, made up of gears, circuits, and transistors. However, we have come to realise that computation is substrate-independent, which means it can be performed using a variety of substrates, including mechanical parts, electronic components, optical systems, quantum states, or even water flowing through pipes. The main point is that the physical substrate is not the key aspect, but rather the abstract structure of information processing that makes computation possible. In a similar vein, spacetime could also be substrate-independent. Think back to where this book began: six numbers. When Mission Control tracked Apollo 8 to the Moon, those six coordinates, three of position, three of velocity, worked because space behaved like a reliable, three-dimensional stage. General relativity then showed that the stage itself was dynamic, curving in response to mass and energy. What the evidence in this book suggests is that the stage is not just dynamic but emergent that those six numbers describe a geometry which crystallises from quantum information, the way a musical performance emerges from the abstract structure of a score, not from the wood of any particular instrument. The important factor might not be the points in geometric space, but rather the patterns of quantum correlation. Geometry serves as a useful descriptive tool and a framework that emerges from certain interactions, making it convenient for specific purposes, but it does not represent fundamental reality. This raises further questions. If quantum states in Hilbert space are fundamental, what about Hilbert space itself? It's an abstract mathematical structure, an infinite-dimensional vector space with specific properties. Does it "exist" in any meaningful sense, or is Hilbert space itself just another convenient description that awaits a deeper foundation? We can only speculate. Perhaps fundamentality stops at quantum information, or perhaps there are further layers that have not been discovered yet. But one thing is certain: the three-dimensional space we perceive is not fundamental, it's emergent. This fact forces us to reconsider what "fundamental" means.

Substance, or Patterns?

The distinction between substance, the fundamental material of reality, and its properties or patterns is a concept in ancient philosophy. Philosophers like Aristotle, with his ideas of matter and form, and Descartes, with his res extensa and res cogitans, have traditionally assumed the existence of an underlying substance, a notion that continues to influence modern discussions of physical properties in spacetime. Emergent spacetime clearly challenges substance metaphysics. When space emerges from entanglement, spacetime points cannot be a substance. Similarly, if particles also emerge from entanglement patterns, it implies that there is no fundamental material substrate. Ultimately, all that remains are patterns, including patterns of quantum correlation, entanglement patterns, and patterns that create the appearance of space, matter, and properties. It has patterns all the way down. Take the example of an electron. Quantum field theory describes it as an excitation of the electron field. But the notion that spacetime is emergent changes this perspective; the electron field is no longer fundamental as it relies on space to exist. The electron then emerges as a pattern within the entanglement network, exhibiting a stable configuration that resembles a particle under specific conditions. But patterns require something to follow a pattern. If an electron is a pattern, then what is it made of? Language fails here. We want to ask what quantum information is made of. But this question assumes a framework of substance and property that may not be relevant. Quantum information is not made of anything more fundamental as it simply exists. Patterns exist mathematically and evolve according to the Schrödinger equation, requiring no substrate beyond Hilbert space itself. Some philosophers find this troubling. They question how patterns can exist without substance, and how structure is possible without something to be structured. Others find it liberating. Perhaps substance was always a confused concept, a category mistake that arises from over-generalising ordinary experience. We are made of atoms, atoms are made of particles, and we naturally assume this hierarchy continues to some truly fundamental substance. But why should ultimate reality conform to categories from our everyday experience? Mathematics provides an analogy. The number three doesn't exist as a substance. It's made up of nothing. Yet three is real - it has properties, relationships to other numbers, and a role in mathematical structures. We can reason about three, prove theorems about it, and use it in calculations. Its reality is structural, rather than substantial. Physics could be entirely like this: reality might be made up of mathematical patterns that satisfy specific consistency requirements, with no underlying substance beyond these patterns. This perspective is known as structural realism, which maintains that reality is composed of structures and relations rather than inherent properties of matter. Emergent spacetime naturally leads us towards structural realism. If space emerges as a pattern, what can be considered substantial? The only persistent aspect is the mathematical structure, comprising Hilbert spaces, quantum states, unitary transformations, and entanglement. These elements dictate how reality unfolds, and they are purely structural, consisting of relationships and patterns rather than physical entities. Does this mean reality is just mathematics? Not quite. Mathematics is abstract and timeless, but physical reality is different. It has specific initial conditions, particular outcomes, and actual events. Our universe is one specific instance in an infinite Hilbert space. The question is, why does this state exist and not another? Mathematics alone cannot answer this question because it cannot explain why our universe is the way it is. The fundamental nature of reality (what actually exists) seems to be a mathematical structure that takes shape in specific configurations and evolves according to mathematical laws. This means we have a pattern without substance, information without a physical base: structure that goes all the way down. If space, particles, and physical properties arise from quantum information, then what happens to consciousness?

Consciousness

The hard problem of consciousness is this: how do physical processes give rise to subjective experience? What is it about being me, seeing red, feeling pain, or having thoughts that makes them feel the way they do? How do objective physical states lead to the subjective experience we have? Emergent spacetime does not provide a solution to this problem; in fact, it makes it even more challenging. We previously believed that consciousness relied on specific arrangements of matter, prompting the question of which arrangements give rise to consciousness and why. However, with matter and space now being emergent, it is clear that consciousness emerges from patterns of quantum information, making the relationship between consciousness and the physical world even more complex. But there's a possible silver lining. One approach to consciousness, developed by neuroscientist Giulio Tononi, proposes that consciousness is integrated information. This framework, known as IIT, remains highly controversial: a 2023 open letter signed by over 120 researchers raised serious concerns, and it is contested by many neuroscientists and philosophers of mind. With that caveat noted, the framework offers one possible lens. Systems that process information in specific ways have varying degrees of consciousness. The more information is integrated when it is processed, the richer the conscious experience becomes. If this approach is correct and reality is fundamentally informational, then consciousness becomes less mysterious. It's not a separate substance that magically arises from matter, but rather a sophisticated form of information processing appearing in systems with sufficient integration and complexity. The human brain has a unique structure, consisting of roughly 86 billion neurons with trillions of connections, which enables it to process information in a highly integrated manner. When viewed from the emergent spacetime perspective, the brain appears as a complex configuration of quantum fields, forming a specific entanglement pattern that gives rise to conscious information processing when observed at the appropriate scales. This does not fully address the difficult problem. We still do not understand why integrated information gives us a sense of experience, or why we have subjective experiences instead of just processing information. However, it categorises consciousness alongside other phenomena that emerge from simpler components. A computer simulation of weather does not contain wind. It contains numbers in silicon, voltages, logic gates, memory addresses. Yet the output is a hurricane. Consciousness may stand in the same relation to quantum fields: not contained in them, but produced by them, at a level of description that cannot be reduced away. Some go further, proposing that consciousness might be fundamental. What if experience is a basic part of quantum information, present at every level, with human consciousness being made up of simpler elements of experience? This is panpsychism: the view that consciousness is everywhere, with complex forms coming from simpler parts. Panpsychism might sound radical, but it is not obviously wrong. If quantum states are fundamental and consciousness is fundamental, perhaps they are two aspects of the same thing. The quantum state of an electron might have a primitive experiential character alongside its physical properties, with human consciousness emerging from the combination of vast numbers of such elementary experiences. I am sceptical of panpsychism for several reasons, mainly because it fails to provide a clear explanation of how basic experiences merge to create complex consciousness. However, its credibility improves slightly if that what we call spacetime arises from something deeper. If space is not a basic entity, then why should consciousness be considered fundamental either? It is possible that both mental and physical aspects arise from quantum information, or alternatively, they might be complementary facets of the same underlying reality that we have not yet fully understood. The view that consciousness is emergent does not diminish its reality or its ability to influence outcomes. Temperature, for instance, is an emergent property of molecular motion, but it can still inflict burns. Likewise, consciousness emerges from the processing of neural information, and it has tangible effects. Your choice to read this book is a case in point: although it can be explained in terms of quantum field configurations, it remains a real decision with real consequences. The compatibilist view makes sense to me because it acknowledges that consciousness is real yet not a fundamental aspect of the universe. Instead, consciousness emerges from the specific ways matter and energy are arranged. These arrangements, in turn, originate from quantum fields, which themselves are a result of complex entanglement patterns. A clear hierarchy of emergence exists, beginning with quantum information and progressing upwards, with consciousness positioned within the intermediate layers of this hierarchy. But I acknowledge the uncertainty that surrounds us. Consciousness remains a deeply puzzling phenomenon. The emergent spacetime theory offers a fresh approach to understanding it, but it has a long way to go before we can be sure if it will unravel the mystery or just redirect our inquiry to another area. If reality is quantum information that evolves according to mathematical laws, where does free will fit in? Traditional debates about free will assume a physical world of particles moving through space, either in a deterministic way (classical physics) or a probabilistic way (quantum mechanics). The question is: are human choices determined by what happened before, or do we really have the freedom to choose? Emergent spacetime completely changes this debate. We are part of the physical world, not separate from it. The patterns in quantum fields make up our brains, and our thoughts process information. Our choices come from how entanglement states evolve. There is no separate decision-maker, no ghost in the machine, that exists outside physical law. This does not necessarily mean that determinism wins. First, quantum mechanics is inherently probabilistic. The Schrödinger equation evolves in a deterministic way, but the outcomes of measurements are random. When you are faced with a decision, your brain is in a quantum superposition of several possibilities. The act of choosing, or measurement, gives one outcome, with its probability determined by the quantum state. You cannot predict what choice you will make in advance, not even in theory, because the randomness is fundamental. This doesn't automatically give us robust free will, though. Random choices are not the same as free choices. Quantum randomness is something that happens on its own, and we cannot control it. Simply replacing determinism with randomness does not obviously make things better. But what if free will isn't about being undetermined? What if it's about being determined by the right things, your values, preferences, and reasoning, instead of external coercion? From this compatibilist perspective, you have free will when your choices come from your own deliberation rather than from outside forces. The fact that your deliberation is based on quantum fields that obey laws does not make it any less yours. What matters is that the process belongs to you, it is the emergent pattern we call your consciousness and agency that makes the decision. Agency is real in the same way temperature is real, not despite being emergent, but through it. When you engage in deliberation, evaluating your options and making a decision, a genuine process unfolds. Your deliberation constitutes a real causal process, where your reasons have a large impact and the decision is truly yours, making your agency a tangible and meaningful aspect of the decision-making process. Emergent spacetime does not make agency illusory, any more than knowing the physics of vibrating strings makes a symphony illusory. A symphony is air pressure variations. It is also Beethoven's Ninth. Both descriptions are complete. Neither cancels the other. They operate at different levels, and both are valid. Some philosophers argue that this trivialises free will, as it suggests that our choices are no more than quantum fields evolving, which raises the question: are we really making choices, or are we just deceiving ourselves? I think this misunderstands emergence. At the quantum level, there are only fields and entanglement. However, at the emergent level, where human cognition operates, genuine choices and agency exist. Both descriptions are correct; they answer different questions at different scales. Think about how we describe a city, it is made up of individual people, buildings, and roads. In theory, we could explain everything that happens in Mumbai using just the physics of atoms and photons. But concepts like traffic, culture, economy, and political tension are not illusions that vanish when we look closely. They are real features of the city that can be accurately described at the level where they occur, even if they are not visible at the atomic level. Similarly, the brain has its own levels of description. The cognitive level, where decisions, intentions, and choices are made, is the right level to talk about agency, whatever the underlying quantum fields are doing. Philosophers still debate free will and consciousness, but emergent spacetime provides a framework that brings together the basic quantum description and human experience. As patterns in quantum fields, we have real agency and choice, which are part of these patterns. This framework helps us understand how our choices and consciousness emerge from quantum interactions.

Free Will

To exist, what does it mean? Tables and chairs occupy space as physical objects. The number three, on the other hand, exists as an abstract entity with distinct properties, mathematical relationships, and a specific role in reasoning. Most people agree that it does not exist in the same way as physical objects, and this difference in existence is what sets them apart. Traditionally, physics draws a clear line: physical objects have a concrete existence in spacetime, whereas mathematical structures are abstract. Reality is seen as physical, with mathematics describing it. But if spacetime is an emergent structure arising from quantum correlations, this line becomes blurred. Spacetime points are not fundamental - they are emergent. The quantum state, which is more fundamental, gives rise to spacetime that emerges. The quantum state is a mathematical object, a vector in Hilbert space. So what is more fundamental: the mathematical Hilbert space or the physical spacetime? If we say the Hilbert space is more fundamental and spacetime is emergent, we are giving mathematical structure a status that was previously reserved for physical objects. If we say spacetime is more real, we are privileging the emergent over the fundamental. One response is to fully embrace ontic structural realism. The only thing that is real is structure. Quantum states are made up of structures, which are mathematical relations. Spacetime emerges from this quantum structure. Tables and chairs are structures that exist within the spacetime that emerges. To "exist" means having a place in the relational network that makes up reality. This dissolves the distinction between abstract and concrete existence. All existence is structural. Some structures are fundamental, such as quantum states and their evolution, while others are emergent, like spacetime, particles, and tables. But emergence does not make things less real; it is just a different kind of reality that is appropriate at different scales. Another response is to distinguish between fundamental and derivative existence without claiming that one is more "real" than the other. Quantum information exists fundamentally, while spacetime exists derivatively. Both are real, but each in its own sense. This can be called existence pluralism, having multiple types of existence rather than a single, univocal category. This is reflected in ordinary language, where we *consider* temperature to exist, despite being merely the average molecular kinetic energy. Similarly, we say computer programs exist, although they are essentially voltage patterns in circuits. Nations, too, are said to exist, even if they are just collective agreements between people and institutions. In each of these cases, the entity that emerges has an existence, but it is different from that of the components that make it up. Perhaps what matters is that existence isn't a simple binary property; it's relational. Things exist relative to their levels of description, scales of observation, and frameworks of inquiry. Electrons exist at the quantum level, tables at the human scale, and galaxies at the cosmic scale. None of these are more real than the others; they are all real in their own contexts. This pluralist view of existence can be unsatisfying. We want to know what exists at the deepest level. The desire for this knowledge may stem from treating existence as a thing, instead of a framework for understanding reality in different ways. Emergent spacetime leads us to this pluralist insight. Space emerges, yet it isn't fundamental. Quantum information is fundamental, but it doesn't exist in space. Both types of existence are real, different, and compatible. Reality is rich enough to accommodate multiple levels and types of existence.

Why does the universe follow laws? Why are there consistent patterns and mathematical regularities that govern how things behave? Traditional answers range from metaphysical necessity, which says laws must be as they are, to divine creation, where God chose these laws, to brute fact, which means laws just are, with no further explanation possible. Modern physics tends towards the third option; we discover laws empirically, use them to make predictions, but do not explain why these laws exist rather than others. Emergent spacetime suggests an alternative: perhaps the physical laws themselves emerge from deeper information-theoretic principles. Take Newton's second law: F = ma. Initially, this seemed like a fundamental law of motion. However, we have since learned that it is emergent and valid only at low speeds and under weak gravity. Special and general relativity are more fundamental, with Newton's law emerging as an approximation. Similarly, in the 19th century, the laws of thermodynamics appeared fundamental. Now, we understand them as emerging from statistical mechanics. The second law, that entropy increases, is not a fundamental dynamical law but a statistical consequence of large systems. If spacetime emerges from quantum information, it is possible that all laws governing spacetime also emerge. Einstein's field equations would then be effective descriptions, valid at large scales, but ultimately derived from principles of quantum information. What might these underlying principles be? Several possibilities exist:

- Unitarity. Quantum evolution preserves information, which is a core principle of quantum mechanics, and this conservation of information has important implications. The requirement that information cannot be destroyed leads to certain inevitable behaviours, and many emergent phenomena may arise from this unitarity.
- Entanglement structure stability. Only specific entanglement patterns remain stable over time, giving rise to regular, law-like behaviour at larger scales. In contrast, unstable patterns decay or fluctuate chaotically, suggesting that the laws we observe are a result of what remains stable.
- Quantum error correction. Systems that utilise quantum error correction are robust against noise and decoherence, which implies that physical laws might represent optimised error-correcting codes. The universe exhibits regular patterns because error correction creates stability, making it a fundamental aspect of the laws we observe.
- Computational complexity. Fundamental limits constrain the efficient

processing of information in quantum systems, and physical laws may reflect these constraints. Certain processes are forbidden because they would exceed these fundamental limits, not due to dynamics, but due to the limitations on computational resources.

These ideas are speculative but can be tested in principle. If laws emerge from information-theoretic principles, we should be able to derive known laws from these principles and predict new phenomena. Researchers are pursuing this and have had some success. If this programme succeeds, the implications would be profound. Laws would not be arbitrary features of reality that we happen to discover. They would be necessary consequences of quantum information dynamics, potentially varying in details across different emergent spacetimes, but constrained by a fundamental information structure. This might finally address the ancient question: why is there something rather than nothing? Perhaps "nothing" is unstable from an information-theoretic perspective. Maybe quantum information naturally organises into patterns, necessarily giving rise to emergent spacetime with law-like behaviour. Existence and law might be inevitable consequences of information logic. But we should be cautious. This vision might be too ambitious. Some laws might emerge while others remain fundamental and unexplained. Perhaps many universes exist with different laws, and we are in one that has no deeper explanation. The question "why these laws?" might have no answer beyond "that's just our universe." What we can say is that emergent spacetime provides a framework for asking whether laws might emerge rather than be axiomatic. This, in itself, is philosophical progress, even if the ultimate answers remain elusive. In our Why Space Has Three Dimensions exploration we effectively bridged the gap between physics and philosophy. By starting with empirical discoveries about black holes, we uncovered philosophical questions about information that warranted further exploration. The mathematical results on holography have sparked important debates about the nature of dimensions, and experiments with quantum simulators have compelled us to reconsider our understanding of what is fundamental. The characteristic interplay between discoveries and theories is what drives fundamental physics forward. On the frontier of physics, new findings do not just reveal fresh insights as they have the capacity to radically alter our conceptual frameworks. Quantum mechanics, for example, has dramatically reshaped our understanding of causality, locality, and determinism. Relativity, too, has profoundly impacted our concepts of time, simultaneity, and space. Today, the concept of emergent spacetime is transforming our perceptions of substance, emergence, and the nature of reality itself.

Where does physics end and philosophy begin? The boundary between them is becoming increasingly unclear. Questions about fundamental existence, the nature of emergence, and the status of physical laws are both empirical and conceptual in nature. We cannot answer them through experiments alone, we also need coherent philosophical frameworks to make sense of our findings. But at the same time, we cannot answer them through philosophy alone empirical evidence from physics is also essential. This is what is known as naturalised philosophy, which involves philosophical inquiry that is connected to science, rather than being separate from it. The questions being asked are philosophical in nature, dealing with meaning, existence, and basic categories, but any answers must fit with our current understanding of physics. Some philosophers resist this integration, seeking an autonomous philosophy that is separate from science. Some physicists also resist, dismissing philosophy as mere speculation. However, emergent spacetime demonstrates why both views are mistaken. Physics gives rise to unavoidable philosophical questions, and to answer them, we need both empirical rigour and conceptual clarity. We will require physicists to think philosophically and philosophers to engage seriously with physics going forward. The questions we face, including what is fundamental, how emergence works, and the meaning of existence, cannot be answered from within traditional disciplinary boundaries. To find the answers, physicists and philosophers can adopt a collaborative approach, showing mutual respect and a willingness to cross these boundaries.

Still Beginning

We have now reached the limit of what current physics and philosophy can tell us. Spacetime comes from quantum information. There are strong reasons to think that particles, forces, and perhaps even the laws of physics emerge the same way. Frameworks for understanding consciousness, free will, and existence within an emergent universe are being developed, though none are yet complete. But questions persist: What, if anything, lies deeper than quantum information? Is there a foundation beneath Hilbert space and unitary evolution, or does the explanation stop here? I want to be precise about what this book has not answered, because the unanswered questions are where the interesting work is. Why quantum mechanics at all? We have used quantum mechanics as the foundation throughout Hilbert spaces, unitary evolution, the Born rule for probabilities but we have not explained why physics is quantum mechanical rather than classical. This is not a gap in the argument; it is a deeper question than the argument addresses. The emergence of geometry from entanglement assumes the entanglement. What produces the entanglement? The hard problem of consciousness sits in a similar position. If spacetime, particles, and forces are all emergent, and if we are physical systems made of these things, then our experience of the world; the redness of red, the felt quality of a thought must also, somehow, be emergent. The philosophical framework for how that happens is contested. I do not think physics has a satisfying answer yet, and I am wary of the tendency to invoke quantum mechanics whenever consciousness becomes difficult. What I can say is that a world in which space itself is not fundamental is a world that demands a rethinking of what kind of thing a conscious observer is. The question is real. The answer is not in this book. I mention it not to gesture vaguely at mystery, but because the reader who has followed the argument this far deserves to know where it ends. Some might be answered by combining theory and experiment. Others might remain permanently beyond our grasp, limited by fundamental constraints on knowledge itself. The nature of inquiry at the frontier is such that each answer we uncover gives rise to new questions, and every mystery we solve reveals even deeper enigmas. The progress in understanding the foundations of space has been real, but it has also been humbling. Our journey so far has led us to a clear realisation: quantum information may be the foundation of reality, an insight that, if confirmed, will reshape physics from its foundations. The path forward requires new experiments that probe emergent spacetime directly, mathematical frameworks that make the emergence of geometry calculable, and philosophical concepts adequate to a reality where space is not given but made. The next revolution might come from quantum computing reaching scales where we can simulate large regions of emergent spacetime. It might come from gravitational wave astronomy detecting signatures of quantum structure. It might come from a theoretical breakthrough that connects quantum information to consciousness. Alternatively, as is often the case with revolutions, it might come from somewhere entirely unexpected. We may be witnessing a turning point in humanity's understanding of reality. If spacetime emerges from quantum information, it will rank among the most impressive discoveries in physics with far-reaching implications that will impact metaphysics, epistemology, and the philosophy of mind for years to come, shaping the way we think about the world and our place in it. Physics is not ending. It is starting over, with harder questions and stranger answers than before. The direction is forward, into territory we cannot yet map, guided by the oldest tools we have: careful observation and clear thought. We are made of the patterns we are trying to understand. That is the deepest mystery of all, and the reason the search continues.

22

SPACE IS A VERB, NOT A NOUN

We began with a simple question: why does space have three dimensions? The answer, which spans twenty-two chapters is this: I think space has three dimensions because quantum information finds, in three, its most stable home. The number three is not a random occurrence or an arbitrary selection, but rather might be a natural consequence of the dynamics of quantum entanglement. This conclusion is both satisfying and unsettling. It is satisfying because it is grounded in mathematics and supported by experimental evidence. It is unsettling because it changes what we thought we knew about space and physics. *Space is not a container, but a consequence.* Spacetime is not the stage where physics takes place; it is what physics creates. Our journey started with Chandrayaan-3, then looking at Earth from space and wondering why that view shows three spatial dimensions. The question appeared simple, perhaps even trivial. It wasn't. Our pursuit of an answer led us into black hole thermodynamics, revealing that information is a fundamental aspect, while geometry may be derivative. We also examined holography, which demonstrated how three-dimensional gravity can emerge from two-dimensional quantum field theory. Furthermore, our investigation of quantum entanglement and tensor networks clarified the mathematical structure underlying emergence. We investigated the limitations of exotic dimensions, revealing that two dimensions are too restrictive and four or more are unstable.

Our exploration of quantum simulators showed that they can create simplified geometries in laboratory settings. By analysing gravitational waves and spacecraft measurements, we tested the notion that spacetime is smooth, or whether it has a quantum structure. We also examined the potential implications of engineering an emergent geometry, considering what this might entail. We also explored the philosophical implications, asking questions like: what is truly fundamental, what does it mean to exist, and how do consciousness and agency emerge in the universe? Throughout, a theme recurred: the information-theoretic perspective transforms everything. Once you view reality as patterns of quantum correlation rather than objects in space, the question shifts from "why three dimensions?" to "why does quantum information organise into three-dimensional geometry?" That question might have an answer.

The Answer

The argument of this book, compressed to its essentials, is this:

Quantum systems could naturally entangle when they interact, developing correlations that classical explanations cannot account for. Entanglement is fundamental and ubiquitous, with information at its core. Specific patterns of entanglement give rise to geometric relationships, and when many qubits entangle in particular ways, their entanglement structure can be viewed as geometry, complete with distances, angles, and dimensions derived from the pattern of quantum correlations. However, not all patterns of entanglement are stable, as most configurations are prone to breaking down, changing chaotically, or becoming inconsistent. Only special patterns endure, namely those that can withstand disturbances and exhibit the regularity observed in physical laws. The most stable general configuration is three-dimensional space, which emerges as the norm when entanglement patterns settle into configurations governed by quantum mechanics and entanglement dynamics. This is because quantum information could naturally form a three-dimensional spatial structure, similar to a river that naturally flows along the path of least resistance in a landscape, not because it was deliberately chosen, but because the laws of physics have determined it. Two dimensions are insufficient, as two-dimensional space lacks the structure necessary for complex entanglement hierarchies, resulting in systems that either expand to higher dimensions or break into one-dimensional pieces. Conversely, four or more dimensions are too permissive, with higher-dimensional spaces having too many connections and too few limits, leading to instability. Entanglement networks that attempt to form four-dimensional geometry typically become unstable or effectively reduce to three dimensions as they evolve. Ultimately, three dimensions are ideal because the dynamics of quantum information favour them, given the laws of quantum mechanics, the nature of entanglement, and the time required for patterns to settle, resulting in three-dimensional space emerging as the typical outcome.

Our universe has three spatial dimensions for a specific reason. It is not because of a divine choice, nor is it necessary for observers to have three dimensions, nor is it a result of random selection from infinite possibilities. Quantum information, following its own principles, leads to the emergence of three-dimensional geometry as the most stable pattern at a large scale, which is why we observe three dimensions in our universe.

What This Means

Thinking of space as emergent rather than fundamental alters our approach to physics. We have viewed space as the setting for physical events for over three centuries, since Newton's time. Objects move through space, fields fill it, and forces act within it. Space was seen as a fixed and unchanging background. Einstein changed this by making spacetime dynamic, showing that it curves when matter and energy are present and expands with the universe. But spacetime is still a basic concept. General relativity explains how spacetime curves, but it does not tell us what spacetime is or why it exists. Now, evidence suggests that spacetime may not be as fundamental as we thought. This does not mean spacetime is not real. Things can be real even if they come from other things. Space is real because you are moving through it, light travels through it, and gravity affects it. It is like a rainbow, which is a real thing that happens when light and water are present, even if it is not a basic part of nature. What is fundamental, then? The basics are quantum information, which includes Hilbert spaces, quantum states, unitary evolution, and entanglement. These concepts are the foundation of every quantum mechanics formulation and cannot be broken down further, representing the deepest level of our current understanding. From this foundation, everything else emerges, such as space from patterns of entanglement, time from how entanglement evolves, particles from stable excitations, forces from interactions in the geometry that emerges, and possibly even physical laws from the need for information to be consistent. Assuming this framework holds true, our reality seems to comprise patterns rooted in quantum correlations. The human body is made up of patterns, our thoughts process information within these patterns, and space itself is a pattern that extends to the Planck scale, where geometry transforms into pure quantum information. This forms the new basis of physics, shifting from spacetime and particles to quantum information and emergence. Our fundamental understanding of the world may be completely overturned; everything we believed to be basic could be emergent, and we now possess a mathematical language to thoroughly examine this concept, including the nature of space, which is contingent, emergent, and explicable through information organisation.

Open Questions

This answer raises new questions, and every answer does the same. The emergence of time is not well understood, and this is a problem we need to solve. Our focus has been on spatial dimensions, but time is just as important. New research suggests that time may arise from entanglement evolution or renormalisation group flow, which is an interesting idea. However, we still have a lot to learn about how time emerges, and we want to understand it as well as we understand spatial emergence. Another question is what determines stability. We know that three dimensions usually emerge, but we do not know why this is the case. Can we find the exact criteria for stability and work them out from basic principles? It is also possible that unusual stable configurations could lead to geometries that are very different from the ones we see in our universe. The relationship between classical General relativity and discrete quantum information is another area we need to explore. We have holographic dualities and tensor network models, but we still need to find the complete path from microscopic to macroscopic. Deriving Einstein's field equations from the dynamics of quantum information would be a major breakthrough. Finally, we have to ask whether particles and forces emerge in the same way as spacetime. We have some evidence and theoretical frameworks that suggest this might be the case, as we saw in Chapter 20, but we are not certain. If we can derive the Standard Model from the structure of entanglement, it would be a big step forward for particle physics. These questions will continue to challenge researchers for decades. Some may yield to a combination of theoretical and experimental approaches. Others might remain unresolved for generations, and a few could be inherently unanswerable due to observational or computational constraints. This is the essence of pioneering science: each breakthrough uncovers new frontiers to investigate. The understanding of spacetime is not a culmination of physics, but a starting point for new inquiry, ushering in a fresh wave of exploration and discovery.

The Path Forward

Now that we have reached this point, what is our next step? Quantum simulation is the most pressing frontier in modern research. The development of larger quantum computers will enable direct testing of theories related to emergent geometry, leading to the discovery of new phenomena that current calculations cannot predict. Every major quantum computing initiative should treat emergence as a core research question. Holographic duality serves as our primary mathematical tool, with AdS/CFT and related correspondences allowing for precise calculations of geometric emergence from boundary quantum theories. These dualities must be extended to realistic cosmologies to further our understanding. Tensor networks provide an explicit means of constructing emergent geometry, with ongoing research focused on improving these methods, determining which networks yield specific geometries, and establishing connections to continuum physics. The study of gravitational wave astronomy may uncover signatures of quantum structure, and as detection technology advances, it will be possible to probe gravity under extreme conditions, potentially revealing emergent structure. The exchange of ideas among different approaches, including holography, loop quantum gravity, causal sets, and string theory, is essential for the advancement of the field. A pluralistic and open-minded approach, rather than a narrow focus, will drive progress. The development of quantum information theory is also necessary, particularly in understanding entanglement in relativistic contexts, identifying geometric patterns, and establishing stability criteria. Ultimately, philosophical clarity is vital to guiding research in this complex and interdisciplinary field. As we reshape our understanding of space, time, and existence, clear thinking is the prerequisite. We must collaborate to drive progress. The landscape is indeed very vibrant, with conferences on emergent spacetime drawing hundreds of researchers and papers being published weekly. New experimental capabilities are emerging every year, and young researchers are joining the field in growing numbers, which suggests that the momentum is gathering pace and is likely to be sustained.

A Timeline

Here's a possible trajectory for the coming decades, grounded in current physics, although it is speculative about the pace. Near term (5-10 years), quantum simulators with thousands of qubits may routinely show emergent geometries. Holographic dualities could be extended to more realistic cosmologies. Tensor network methods might become standard tools in quantum gravity. We may see hints of how particles emerge with space. Medium term (10-30 years), we may develop quantum computers that can simulate Planck-scale physics in small regions, allowing us to conduct direct experimental tests of emergence mechanisms. We might be able to derive Standard Model symmetries from information-theoretic principles. Gravitational waves or cosmological data could show subtle signs of emergent structure. A significant theoretical breakthrough might link quantum error correction to physical law. Long term (30-100 years), we may achieve a full understanding of how classical spacetime emerges from quantum information and derive both general relativity and the Standard Model from entanglement dynamics. It could become possible to experimentally manipulate local geometry through vacuum engineering. We may create artificial regions with tailored properties. Our understanding might also extend to the emergence of consciousness, leading to a synthesis as consequential as quantum mechanics or relativity were for the 20th century. Very long term (100+ years), spacetime engineering technologies might become practical. A deeper understanding might reveal that quantum mechanics itself emerges from something more fundamental. We could address some profound questions about the origins of mathematics, physical laws, and parallel universes. Physics might shift from studying laws in spacetime to understanding how both laws and geometry emerge from information. This timeline may prove overly ambitious, with progress unfolding at a slower pace, obstacles appearing more formidable, and challenges becoming more daunting than anticipated. Alternatively, it could be too cautious, and unprecedented breakthroughs might accelerate the discovery process. One thing is certain; we have reached a critical juncture. Recognising spacetime as emergent rewires our approach. The questions we are now able to pose, the experiments we can devise, and the theories we can formulate are radically different from those of just two decades ago.

What it Means to Inhabit Space?

To end, I would like to talk about something that goes beyond the technical side of physics. Modern physics has dissolved the boundaries between humanity and nature progressively. We're made up of atoms that were forged in stars. Our brains are subject to the same laws as everything else in the physical world. We are not separate observers. We are a part of it. If space emerges from quantum information, this unity becomes deeper. We inhabit space, but we are also spatial patterns emerging from the same informational substrate. The boundary between the observer and the observed becomes even less clear. As you read these words, patterns of neural activity in quantum fields shape your thoughts. The space that contains your mind emerges from entanglement. Light carries information from the page to your eye, propagating through a geometry that arises from underlying systems. The entire scene, including the reader, words, and the space between them, consists of beautiful, intricate patterns within patterns, ultimately rooted in quantum information. *We are the universe, and it is understanding itself through us.* The patterns that make us have evolved to recognise who we are, to examine the structures that make up our existence, and to grasp the mathematics that underlie everything around us. I believe this makes life more meaningful. It shows we are not just random accidents in an indifferent universe. Instead, we are the result of quantum information's capabilities, with patterns evolving into conscious beings that think and understand. Fundamental physics and human consciousness share the same underlying principles. We are an integral part of these principles, and our existence is the most remarkable expression of them, the moment when a pattern becomes aware of itself. That's not reductionism; that's wonder.

The Question Persists

We have reached the end of this particular search. The inquiry itself does not end. As the popular saying goes, each conclusion we draw marks the beginning of a new investigation, and every answer we find has more questions following closely behind. Why does space have three dimensions? Because three might only be the dimensionality that survives the intersection of every constraint that quantum mechanics, information theory, and gravitational stability impose. This answer emerged after covering twenty-two chapters of physics, mathematics, and philosophy. But this answer raises more questions. Why does quantum information exist? Why are these mathematical laws governing it? Why is there something instead of nothing? Are there other universes with different patterns? Is there a level more fundamental than quantum information? The answers to these questions have not been found yet. Future generations of physicists and philosophers will uncover them. As each answer emerges, it will lead to new questions, either expanding our investigation or revealing that some things exist without requiring further explanation. For sure, we have made thrilling progress, and our achievements seem clear. The puzzle of three dimensions, once deemed impossible even to approach, now has a compelling candidate answer, built on rigorous mathematics and the first experimental tests of its foundations. The power of science lies in its ability to ask questions about the world, develop theories to explain them, test these theories against reality, and ultimately achieve a deeper understanding that transforms our perspective on the world. We exist in a three-dimensional space that emerges from quantum entanglement. This space is not a passive container, but an active outcome: a complex pattern that forms when a deeper structure crystallises, giving rise to a specific geometry as quantum information organises itself in distinct ways, ultimately shaping our reality. As intelligent beings with complex patterns, we have the ability to ask why. And this question of why we have three dimensions has brought us to this point, changing our understanding of physics, philosophy, and our position in the universe. Yet, this journey continues. It carries us forward into deeper questions and astonishing truths. The last page of this book is simply the first page of our next chapter, as humanity strives to understand the patterns we embody and the realities we inhabit.

Space is a verb, not a noun. We are what the cosmos achieves when it wakens to itself.

And so, the awe persists.

Notes

Science has never been more accessible, and that is largely thanks to a number of institutions and publications that have made it their mission to bring serious ideas to a general audience without condescending to the quality. What follows is a personal list of some of my favourite sources.

- Quanta Magazine: The gold standard of science journalism, Quanta delivers in-depth coverage of mathematics, physics, and biology, thanks to reporters who genuinely understand the science and editors who are committed to accuracy. Say when a breakthrough in quantum gravity or cosmology emerges from the arXiv and enters public discourse, Quanta is usually the best place to learn about it first, providing readers with authoritative insights and explanations.
- The Royal Institution: The Royal Institution, founded in 1799, continues to thrive. The RI's Christmas Lectures have got a long history of inspiring young people to explore science, and their extensive YouTube archive, which now boasts hundreds of talks, showcases some of the most brilliant scientific minds of our time sharing their knowledge with the general public. What is particularly noteworthy is that all of this content is available completely free of charge.
- International Centre for Theoretical Sciences, Bengaluru: ICTS stands out as one of the world's premier centres for theoretical physics, nestled in the hills outside Bengaluru. The organisation runs schools and workshops that attract researchers from all over the globe, and they have got a notable proportion of their work available online.
- Institute for Advanced Study, Princeton: The IAS has been the catalyst for some of the most pioneering work in twentieth-century physics,

and its public lecture series and online videos offer a fascinating glimpse into this rich legacy.

Notes

1. Abbott, Edwin A. *Flatland: A Romance of Many Dimensions.* Dover, 1992.
2. Acharya, Rajeev, et al. "Quantum Error Correction below the Surface Code Threshold." Nature 638 (2025): 920–926.
3. Almheiri, Ahmed, Xi Dong, and Daniel Harlow. "Bulk Locality and Quantum Error Correction in AdS/CFT." Journal of High Energy Physics 04 (2015): 163.
4. Almheiri, Ahmed, Netta Engelhardt, Donald Marolf, and Henry Maxfield. "The Entropy of Hawking Radiation." Reviews of Modern Physics 93 (2021): 035002.
5. Almheiri, Ahmed, Raghu Mahajan, Juan Maldacena, and Ying Zhao. "The Entropy of Bulk Quantum Fields and the Entanglement Wedge of an Evaporating Black Hole." Journal of High Energy Physics 12 (2019): 063.
6. Almheiri, Ahmed, Donald Marolf, Joseph Polchinski, and James Sully. "Black Holes: Complementarity or Firewalls?" Journal of High Energy Physics 02 (2013): 062.
7. Amelino-Camelia, Giovanni, et al. "Quantum-Spacetime Phenomenology." Living Reviews in Relativity 16 (2013): 5.
8. Ambjorn, Jan, Jerzy Jurkiewicz, and Renata Loll. "The Spectral Dimension of the Universe Is Scale Dependent." Physical Review Letters 95 (2005): 171301.
9. Ananthaswamy, Anil. *The Edge of Physics: A Journey to Earth's Extremes to Unlock the Secrets of the Universe.* Houghton Mifflin Harcourt, 2010.
10. Ashby, Neil. "Relativity in the Global Positioning System." Living Reviews in Relativity 6 (2003): 1.
11. Ashtekar, Abhay. "New Variables for Classical and Quantum Gravity." Physical Review Letters 57 (1986): 2244.

12. Bekenstein, Jacob D. "Black Holes and Entropy." Physical Review D 7 (1973): 2333.
13. Bell, John S. "On the Einstein Podolsky Rosen Paradox." Physics 1 (1964): 195.
14. Bertrand, Joseph. "Théorème relatif au mouvement d'un point attiré vers un centre fixe." Comptes Rendus 77 (1873): 849.
15. Bhattacharyya, Sayantani, Veronika E. Hubeny, Shiraz Minwalla, and Mukund Rangamani. "Nonlinear Fluid Dynamics from Gravity." Journal of High Energy Physics 02 (2008): 045.
16. Bousso, Raphael. "The Holographic Principle." Reviews of Modern Physics 74 (2002): 825.
17. Braunstein, Samuel L., and Arun Kumar Pati. "Quantum No-Hiding Theorem." Physical Review Letters 98 (2007): 080502.
18. Brydges, Tiff, et al. "Probing Rényi Entanglement Entropy via Randomized Measurements." Science 364 (2019): 260.
19. Cao, ChunJun, and Sean M. Carroll. "Bulk Entanglement Gravity without a Boundary." Physical Review D 97 (2018): 086003.
20. Cao, ChunJun, Sean M. Carroll, and Spyridon Michalakis. "Space from Hilbert Space: Recovering Geometry from Bulk Entanglement." Physical Review D 95 (2017): 024031.
21. Carleo, Giuseppe, and Matthias Troyer. "Solving the Quantum Many-Body Problem with Artificial Neural Networks." Science 355 (2017): 602–606.
22. Carlip, Steve. "Dimension and Dimensional Reduction in Quantum Gravity." Classical and Quantum Gravity 34 (2017): 193001.
23. Carroll, Sean. *The Biggest Ideas in the Universe: Space, Time, and Motion.* Dutton, 2022.
24. Carroll, Sean. *Something Deeply Hidden: Quantum Worlds and the Emergence of Spacetime.* Dutton, 2019.
25. Carter, Brandon. "Large Number Coincidences and the Anthropic Principle in Cosmology." In IAU Symposium 63, edited by Malcolm S. Longair, 291–298. Reidel, 1974.
26. Chalmers, David. *The Conscious Mind: In Search of a Fundamental*

Theory. Oxford University Press, 1996.

27. Chou, Aaron S., et al. "MHz Gravitational Wave Constraints with Decameter Michelson Interferometers." Physical Review D 95 (2017): 063002.
28. Coffman, Valerie, Joydip Kundu, and William K. Wootters. "Distributed Entanglement." Physical Review A 61 (2000): 052306.
29. Connes, Alain, and Carlo Rovelli. "Von Neumann Algebra Automorphisms and Time-Thermodynamics Relation in Generally Covariant Quantum Theories." Classical and Quantum Gravity 11 (1994): 2899.
30. Corboz, Philippe, Román Orús, Bela Bauer, and Guifré Vidal. "Simulation of Strongly Correlated Fermions in Two Spatial Dimensions with Fermionic Projected Entangled-Pair States." Physical Review B 81 (2010): 165104.
31. DeWitt, Bryce S. "Quantum Theory of Gravity. I. The Canonical Theory." Physical Review 160 (1967): 1113.
32. Dicke, Robert H. "Dirac's Cosmology and Mach's Principle." Nature 192 (1961): 440–441.
33. Ehrenfest, Paul. "In What Way Does It Become Manifest in the Fundamental Laws of Physics That Space Has Three Dimensions?" Proceedings of the Amsterdam Academy 20 (1917): 200–209.
34. Einstein, Albert. "Erklärung der Perihelbewegung des Merkur aus der allgemeinen Relativitätstheorie." Sitzungsberichte der Preussischen Akademie der Wissenschaften (1915): 831–839.
35. Einstein, Albert, Boris Podolsky, and Nathan Rosen. "Can Quantum-Mechanical Description of Physical Reality Be Considered Complete?" Physical Review 47 (1935): 777.
36. Engelhardt, Netta, and Aron C. Wall. "Quantum Extremal Surfaces: Holographic Entanglement Entropy beyond the Classical Regime." Journal of High Energy Physics 01 (2015): 073.
37. Evenbly, Glen, and Guifré Vidal. "Tensor Network Renormalization." Physical Review Letters 115 (2015): 180405.
38. Everitt, C. W. Francis, et al. "Gravity Probe B: Final Results of a Space Experiment to Test General Relativity." Physical Review Letters 106

(2011): 221101.

39. Feynman, Richard P. "Simulating Physics with Computers." International Journal of Theoretical Physics 21 (1982): 467–488.
40. Fishman, Matthew, Steven R. White, and E. Miles Stoudenmire. "The ITensor Software Library for Tensor Network Calculations." SciPost Physics Codebases (2022): 4.
41. Georgescu, Iulia M., Sahel Ashhab, and Franco Nori. "Quantum Simulation." Reviews of Modern Physics 86 (2014): 153.
42. Gu, Zheng-Cheng, Michael Levin, and Xiao-Gang Wen. "Tensor-Entanglement Renormalization Group Approach as a Unified Method for Symmetry Breaking and Topological Phase Transitions." Physical Review B 78 (2008): 205116.
43. Hartle, James B., and Stephen W. Hawking. "Wave Function of the Universe." Physical Review D 28 (1983): 2960.
44. Hastings, Matthew B. "An Area Law for One-Dimensional Quantum Systems." Journal of Statistical Mechanics (2007): P08024.
45. Hawking, Stephen W. "Black Hole Explosions?" Nature 248 (1974): 30–31.
46. Hensen, Bas, et al. "Loophole-Free Bell Inequality Violation Using Electron Spins Separated by 1.3 Kilometres." Nature 526 (2015): 682.
47. Hubeny, Veronika E., Mukund Rangamani, and Tadashi Takayanagi. "A Covariant Holographic Entanglement Entropy Proposal." Journal of High Energy Physics 07 (2007): 062.
48. Hulse, Russell A., and Joseph H. Taylor. "Discovery of a Pulsar in a Binary System." Astrophysical Journal 195 (1975): L51.
49. ISRO. *Chandrayaan-3 Mission: Science Results from the Vikram Lander and Pragyan Rover.* Indian Space Research Organisation Technical Report, 2023.
50. Isham, Christopher J. "Canonical Quantum Gravity and the Problem of Time." arXiv:gr-qc/9210011 (1992).
51. Islam, Rajibul, Ruichao Ma, Philipp M. Preiss, M. Eric Tai, Alexander Lukin, Matthew Rispoli, and Markus Greiner. "Measuring Entanglement Entropy in a Quantum Many-Body System." Nature 528 (2015):

77–80.

52. Jacobson, Ted. "Thermodynamics of Spacetime: The Einstein Equation of State." Physical Review Letters 75 (1995): 1260.
53. Jafferis, Daniel, et al. "Traversable Wormhole Dynamics on a Quantum Processor." Nature 612 (2022): 51–55.
54. Kaluza, Theodor. "Zum Unitätsproblem der Physik." Sitzungsberichte der Preussischen Akademie der Wissenschaften (1921): 966–972.
55. Kennefick, Daniel. *No Shadow of a Doubt: The 1919 Eclipse That Confirmed Einstein's Theory of Relativity.* Princeton University Press, 2019.
56. Klein, Oskar. "Quantentheorie und fünfdimensionale Relativitätstheo rie." Zeitschrift für Physik 37 (1926): 895–906.
57. Ladyman, James, and Don Ross. *Every Thing Must Go: Metaphysics Naturalized.* Oxford University Press, 2007.
58. Leslie, John. *Universes.* Routledge, 1989.
59. Levin, Michael, and Xiao-Gang Wen. "Colloquium: Photons and Electrons as Emergent Phenomena." Reviews of Modern Physics 77 (2005): 871.
60. LIGO Scientific Collaboration and Virgo Collaboration. "Observation of Gravitational Waves from a Binary Black Hole Merger." Physical Review Letters 116 (2016): 061102.
61. LIGO Scientific Collaboration and Virgo Collaboration. "Tests of General Relativity with GW150914." Physical Review Letters 116 (2016): 221101.
62. Maggiore, Michele, et al. "Science Case for the Einstein Telescope." Journal of Cosmology and Astroparticle Physics 03 (2020): 050.
63. Maldacena, Juan M. "The Large N Limit of Superconformal Field Theories and Supergravity." Advances in Theoretical and Mathematical Physics 2 (1998): 231.
64. Maldacena, Juan, and Leonard Susskind. "Cool Horizons for Entangled Black Holes." Fortschritte der Physik 61 (2013): 781.
65. MICROSCOPE Collaboration; Touboul, Pierre, et al. "MICROSCOPE Mission: Final Results of the Test of the Equivalence Principle."

Physical Review Letters 129 (2022): 121102.

66. NANOGrav Collaboration. "The NANOGrav 15-Year Data Set: Evidence for a Gravitational-Wave Background." Astrophysical Journal Letters 951 (2023): L8.
67. Nielsen, Michael A., and Isaac L. Chuang. *Quantum Computation and Quantum Information.* Cambridge University Press, 2000.
68. Orús, Román. "A Practical Introduction to Tensor Networks." Annals of Physics 349 (2014): 117.
69. Page, Don N. "Information in Black Hole Radiation." Physical Review Letters 71 (1993): 3743.
70. Page, Don N., and William K. Wootters. "Evolution without Evolution: Dynamics Described by Stationary Observables." Physical Review D 27 (1983): 2885.
71. Pastawski, Fernando, Beni Yoshida, Daniel Harlow, and John Preskill. "Holographic Quantum Error-Correcting Codes: Toy Models for the Bulk/Boundary Correspondence." Journal of High Energy Physics 06 (2015): 149.
72. Penrose, Roger. *The Road to Reality: A Complete Guide to the Laws of the Universe.* Jonathan Cape, 2004.
73. Rangamani, Mukund, and Tadashi Takayanagi. *Holographic Entanglement Entropy.* Springer, 2017.
74. Rovelli, Carlo. *The Order of Time.* Riverhead, 2018.
75. Rovelli, Carlo. *Quantum Gravity.* Cambridge University Press, 2004.
76. Rovelli, Carlo. *Reality Is Not What It Seems: The Journey to Quantum Gravity.* Riverhead, 2017.
77. Ryu, Shinsei, and Tadashi Takayanagi. "Holographic Derivation of Entanglement Entropy from AdS/CFT." Physical Review Letters 96 (2006): 181602.
78. Samal, Jharana Rani, Ravi Gupta, Arun Kumar Pati, and Anil Kumar. "Experimental Test of the Quantum No-Hiding Theorem." Physical Review Letters 106 (2011): 080401.
79. Sharir, Or, Yoav Levine, Noam Wies, Giuseppe Carleo, and Amnon Shashua. "Deep Autoregressive Models for the Efficient Variational

Simulation of Many-Body Quantum Systems." Physical Review Letters 124 (2020): 020503.

80. Steinhauer, Jeff. "Observation of Quantum Hawking Radiation and Its Entanglement in an Analogue Black Hole." Nature Physics 12 (2016): 959–965.
81. Susskind, Leonard. "The Anthropic Landscape of String Theory." In Universe or Multiverse?, edited by Bernard Carr, 247–266. Cambridge University Press, 2007.
82. Susskind, Leonard. *The Black Hole War: My Battle with Stephen Hawking to Make the World Safe for Quantum Mechanics.* Little, Brown, 2008.
83. Susskind, Leonard. "The World as a Hologram." Journal of Mathematical Physics 36 (1995): 6377.
84. Swingle, Brian. "Entanglement Renormalization and Holography." Physical Review D 86 (2012): 065007.
85. Tegmark, Max. *Our Mathematical Universe.* Knopf, 2014.
86. Tegmark, Max. "On the Dimensionality of Spacetime." Classical and Quantum Gravity 14 (1997): L69.
87. 't Hooft, Gerard. "Dimensional Reduction in Quantum Gravity." arXiv:gr-qc/9310026 (1993).
88. Tononi, Giulio. "An Information Integration Theory of Consciousness." BMC Neuroscience 5 (2004): 42.
89. Unruh, William G. "Experimental Black-Hole Evaporation?" Physical Review Letters 46 (1981): 1351.
90. Van Raamsdonk, Mark. "Building Up Spacetime with Quantum Entanglement." General Relativity and Gravitation 42 (2010): 2323.
91. Vasileiou, Vlasios, et al. "Constraints on Lorentz Invariance Violation from Fermi-LAT Observations of Gamma-Ray Bursts." Physical Review D 87 (2013): 122001.
92. Verlinde, Erik. "On the Origin of Gravity and the Laws of Newton." Journal of High Energy Physics 04 (2011): 029.
93. Verstraete, Frank, J. Ignacio Cirac, and José I. Latorre. "Quantum Circuits for Strongly Correlated Quantum Systems." Physical Review A 79 (2009): 032316.

94. Vidal, Guifré. "Entanglement Renormalization." Physical Review Letters 99 (2007): 220405.
95. Weinberg, Steven. "Anthropic Bound on the Cosmological Constant." Physical Review Letters 59 (1987): 2607.
96. Wheeler, John A. "Information, Physics, Quantum: The Search for Links." In Complexity, Entropy, and the Physics of Information. Addison-Wesley, 1990.
97. Will, Clifford M. "The Confrontation between General Relativity and Experiment." Living Reviews in Relativity 17 (2014): 4.
98. Williams, James G., Slava G. Turyshev, and Dale H. Boggs. "Progress in Lunar Laser Ranging Tests of Relativistic Gravity." Physical Review Letters 93 (2004): 261101.

About the Author

Anvay Sheolikar (often loved to be called "Human from Earth!") is a 19-year-old Aerospace Engineering student at Queen's University Belfast, originally from Pune, India. A defense and law enthusiast, he recently proposed India's first orbital defense doctrine, which received recognition from the Ministry of Defence and Government of India. Born in a multi-generational music family, his grandfather, a violinist, was awarded the Sangeet Natak Akademi Amrit Puraskar for his contribution to Indian classical music from the Vice-President of India.

You can connect with me on:

https://x.com/anvay_sheolikar

www.ingramcontent.com/pod-product-compliance
Lightning Source LLC
LaVergne TN
LVHW100252210926
848480LV00003B/908
* 9 7 9 8 9 0 3 9 2 1 7 7 5 *